Reasoning and Formal Logic

Essays on Logic as the Art of Reasoning Well

Richard L. Epstein

Advanced Reasoning Forum

For more information contact:
  Advanced Reasoning Forum
  P. O. Box 635
  Socorro, NM 87801   USA
  www.ARFbooks.org

paperback ISBN 978-1-938421-03-7

e-book ISBN 978-1-938421-04-4

Reasoning and Formal Logic

Essays on Logic as the Art of Reasoning Well

Richard L. Epstein

Essays on Logic as the Art of Reasoning Well

Dedicated to

Henrique Antunes Almeida

with gratitude for
his criticism and encouragement

Preface

This series of books presents the fundamentals of logic in a style accessible to both students and scholars. The text of each essay presents a story, the main line of development of the ideas, while the notes and appendices place the research within a larger scholarly context.

There is no best order in which to read these essays. They overlap, creating a unified view of formal logic, with the discussions in each adding to the others. Some require a background in formal logic, while others set out (briefly) what is needed. References for which no author is specified are for works by me.

The basic theme here is the analysis of formal logic in terms of what metaphysical assumptions we need when we develop the formal systems we use. In my work I try to set out minimal metaphysics that almost all of us can or do share and which we can use as the basis for a variety of formal systems.

In "Valid Inferences and Possibilities" I show how our metaphysics and our notion of logical possibility are intertwined, for what is possible is what entails no contradiction, while what entails no contradiction is what is possible. Formal logics are a way to distentangle these notions.

Particular formal systems can be understood as arising by factoring in further metaphysical assumptions about truth and what we are paying attention to in our reasoning beyond the minimal ones we need to start any investigation of vaild inferences. The structural analysis of that overview is given in "A General Framework for Semantics for Propositional Logics," and the conceptual basis is presented in "Why Are There So Many Logics?" That view is based on a generous conception of truth and its role in reasoning, which I set out in "Truth and Reasoning."

One way we can compare formal systems based on different semantic assumptions is by attempting to translate from one to another, which is the topic of "On Translations."

The essay "Reflections on Temporal and Modal Logic" is a cautionary tale, showing how formal systems built without careful attention to the metaphysics and foundations of the work can lead to nonsense.

The metaphysical assumptions of classical predicate logic make it difficult if not impossible to formalize reasoning that takes account of existence, time, and change, as I explain in "The Timelessness of Classical Predicate Logic." Taking events as the fundamental entities about which we reason in predicate logic has been proposed as a way to overcome that kind of limitation. But I show in "Events in the Metaphysics of Predicate Logic" that doing so complicates without illuminating.

It's long been thought that metaphysical assumptions about the nature of collections and infinite collections are needed in order to characterize all truths of arithmetic. In "Categoricity with Minimal Metaphysics" I show that no such assumptions are needed: a mechanism for quantifying over names, using no more metaphysical assumptions than those of ordinary first-order predicate logic, is enough.

In "Reflections of Gödel's Theorems" I point out that modern discussions and analyses of Kurt Gödel's work often assume Gödel's platonist views. I offer a different perspective, and in doing so I try to place Gödel's work in the larger tradition of logic.

"On the Error in Frege's Proof that Names Denote" is another cautionary tale, showing how important it is to distinguish semantics and syntax in the development of a formal system.

These essays together give a perspective of formal logic as part of the art of reasoning well. In the Postscript to this volume I try to summarize that conception of logic. It is a view that I hope will illuminate and prove useful.

Richard L. "Arf" Epstein
Dogshine, New Mexico, 2015

Publishing history of the essays in this volume

A version of "Possibilities and Valid Inferences" first appeared under the title "Valid Inferences" in *Logic without Frontiers: Festschrift for Walter Alexandre Carnielli on the Occasion of his 60th Birthday*, eds. J.-Y. Béziau and M. E. Coniglio, College Publications, pp. 105–12, 2012. I am grateful to William S. Robinson and Fred Kroon for their comments on that paper and to Henrique Antunes Almeida and Lynn Diane Beene for criticism that helped me to revise it here.

"A General Framework for Semantics for Propositional Logics" was first published in *Methods and Applications of Mathematical Logic*, edited by Walter A. Carnielli and Luiz P. de Alcantara, *Contemporary Mathematics*, vol. 69, American Mathematical Society, 1988, pp. 149–168. A revision of it appeared as Chapter IV of *Propositional Logics*, 1990, from which this further revision is made. I am grateful to Walter A. Carnielli for his help in preparing the original paper, and to João Marcos and Henrique Antunes Almeida for their suggestions for this version.

This discussion of truth and propositions in "Why Are There So Many Logics?" first appeared as Chapter XI of the first edition of *Propositional Logics* in 1990. Only the addition of some footnotes is new here. It is my earliest attempt to sort out some of the ideas presented in the essay "Truth and Reasoning" in this volume, yet it supplements that essay, too. The latter previously appeared in 2013 in *Prescriptive Reasoning* in this series .

This is the first publication of "On Translations." I am grateful to Walter Carnielli, João Marcos, and Henrique Antunes Almeida for many suggestions that improved it.

"Reflections on Temporal and Modal Logic" was written at Dogshine, New Mexico from October, 2012 to March, 2013. I am grateful to Walter Carnielli, Esperanza Buitrago-Díaz, and Henrique Antunes Almeida, and an anonymous referee for suggestions for improving it. It first appeared in *Logic and Logical Philosohpy*, DOI 10.12775/LLP.2014.020

"The Timelessness of Classical Predicate Logic" served as the introduction to the study of time in a typescript *The Internal Structure of Predicates and Names with an Analysis of Reasoning about Process*, which I made available on the Internet in 2010. This is its first publication. I am grateful to Fred Kroon and Henrique Antunes Almeida for suggestions for improving it. "Events as a Foundation for Predicate Logic" started as an appendix to that same draft in 2010. This is its first publication..

The formal work in "Categoricity wtih Minimal Metaphysics" appeared first in my *Classical Mathematical Logic* in 2006. I am grateful to Henrique Antunes Almeida for suggestions for this version.

"Reflections on Gödel's Theorems" originated as a comment circulated among members of the Advanced Reasoning Forum on a draft of the paper by Stanisław Krajewski referred to in that article. This is its first publication. The analysis there of the argument about predicates being linguistic was first presented in Chapter IV.D.4 of *Predicate Logic*.

This is the first publication of "On the Error in Frege's Proof that Names Denote." Except for the last sentence and quote, it was written in 1983.

The Postscript grew out of a talk called "The Scope of Logic" that I gave in Brazil in 2014. I am grateful to HenriqueAntunes Almeida for helping me to see what I was trying to say.

Valid Inferences and Possibilities

What is possible determines what counts as one claim following from another, yet we understand what is possible by investigating when one claim follows from another. Formal logics can help us break out of this circle by reducing possibilities to simpler notions.

Claims and inferences

All our reasoning involves claims and inferences.

Claims and inferences A *claim* is a written or uttered piece of language that we agree to view as being either true or false but not both.

    An *inference* is a collection of claims, one of which is designated the *conclusion* and the others the *premises*, that is intended by the person who sets it out either as showing that the conclusion follows from the premises or investigating whether that is the case.

Some say that what is true or false are abstract objects and that claims are only representatives of those. Inferences, too, they take to be abstract. But, as I emphasize in "Truth and Reasoning" in this volume, they and we use claims and inferences as defined here in reasoning together. From the perspective of those who take logic to be independent of human reasoning, this essay is not about logic and possibility but how we can or should use logic and possibility in our reasoning. We use inferences in at least six ways: arguments, explanations, mathematical proofs, conditional inferences, causal inferences, and prescriptive reasoning. Though the criteria for what qualifies as a good inference depends on which of those we are considering, the following classifications are central in all evaluations.

Valid and strong inferences   An inference is *valid* if it is impossible for the premises to be true and conclusion false at the same time and in the same way.

Invalid inferences are classified on a *scale* from strong to weak. An inference is *strong* if it is unlikely for the premises to be true and conclusion false at the same time and in the same way; it is *weak* if it is not valid or strong.

With a valid inference if the premises are true, the conclusion is true; with a strong inference if the premises are true, then the conclusion most probably is true.[1]

Example 1   Felix teaches at a university.
Therefore, Felix is a professor.

Analysis   Is this inference valid? Strong? Could it be that Felix teaches at a university but isn't a professor? Yes, he could be a temporary instructor or a teaching assistant. The inference isn't valid. It isn't even strong since we can't rule out those possibilities as unlikely.

As the example illustrates, an inference is either valid or it isn't; there are no degrees to it. A single example of a way the premises could be true and conclusion false suffices to show that the inference is not valid. In contrast, inferences are more or less strong, and that depends on our intersubjective evaluation about how likely certain possibilities are. Attempts to substitute objective criteria in terms of probabilities have been unsuccessful, and it is a serious issue whether we can or should rely on any strong argument as showing that the conclusion follows.[2] Here we'll consider evaluations of inferences solely in terms of validity, for that is the scope of formal logic.

Possibilities

To evaluate an inference we talk about ways the world could be. In our reasoning, however, all we ever use are descriptions. This is what we did in Example 1, where I gave a description of how the world could be in which the premise is true and conclusion false: Felix is a teacher and Felix is a temporary instructor.

Example 2 This is a rock. Therefore, it is solid.

Analysis Professor Zzzyzzx says this is not valid. He knows this, he says, by an intuition, a sense of the nature of the world. He says that he cannot communicate this great insight, but he knows that the inference is not valid.

Perhaps Professor Zzzyzzx is right. But we do not accept his evaluation, nor can we reason with him. Inchoate insights or mystical perceptions cannot be accepted if we wish to reason together, for reasoning together requires communication, and what is inchoate cannot, by its nature, be communicated. Or if it can be communicated, as the Zen master or the Christian mystic hopes, it cannot be done in a way that is observable to all.

To invoke a way the world could be, a possibility, we have no choice but to use a description when we wish to reason together. A description of the world is a collection of claims: we suppose that this, and that, and this are true. We do not require that we give a complete description of the world, for no one is capable of presenting such a description nor is anyone capable of understanding one if presented. By using collections of claims to stand in for or as descriptions of possibilities, we need not commit ourselves to a possibility being something real, such as a world in which I am not bald. A determinist can say that invoking possibilities is just a way to factor into our reasoning our ignorance of how the world is.

But what qualifies some collections of claims as describing a possibility and others not? What do we mean by saying that a dog giving birth to a donkey is a possibility, but a square circle is not? If you say that a dog giving birth to a donkey is not a possibility, how are we to decide if you are right?

Perhaps we have different ideas about what is possible. You might consider that it is physically impossible for a dog to give birth to a donkey, knowing all we now know about the biology of these animals. I might say that it is possible; we just don't know how yet.

Or I might say that it is not possible for a dog to give birth to a donkey by any ethical means, whereas you might say that it would be perfectly acceptable morally to interfere with the biology of dogs and donkeys to bring that about.

There are many different notions of possibility: physical possibility, moral possibility, possibility given what has happened up to this time, possibility given what we know up to this time, ... Regardless of which of these we are employing, we always seem to agree that a description of a way the world could be must at least be consistent. That is, it cannot have or entail a contradiction. It must be *logically possible*.

This seems to be the ground from which we all start in our reasoning. What is possible must be consistent. So there is no way the world could be in which there is a square circle. But there seems to be no contradiction inherent in postulating that a dog could give birth to a donkey: it is logically possible.[3]

It might seem we have made some progress in analyzing valid inferences. But the progress is illusory. What is logically possible is what contains or entails no contradiction. But that requires knowing what it means for a collection of claims to entail another claim, which is what we are trying to understand.

What is logically possible depends on how we understand valid inferences. But what is a valid inference depends on how we understand possibilities, particularly logical possibilities. We seem to be in a circle with no way out.

Logic and logical possibilities

One way we could extricate ourselves from the circle is to agree that we will use our informal, intuitive reasoning to determine what is logically possible. But that is to deny the entire project of trying to understand valid inferences, for we would be trying to make explicit a concept that depends on our not making explicit our most fundamental assumptions.

What we can do is investigate parts of our reasoning, picking out just this or that kind of reasoning relative to restricted semantic assumptions that allow for clarity of analysis, calling that "a logic." Then we can have a clearer notion of possibility and of valid inference for that kind of reasoning. As we extend our investigations to allow for more kinds of reasoning, we will have fuller analyses of logical possibilities and valid inferences. But unless we should ever formalize all of reasoning, which seems very unlikely, we shall never have a complete analysis of logical possibilities and of valid inferences. We shall have analyzed at best only valid inferences relative to the specific semantic assumptions we have made, valid inferences for this logic. Still, this is a useful goal.

Semantic reductions

Can we find some properties of claims that might be simpler and in terms of which we could understand possibilities?

We have one property of claims that we have accepted as clear or at least so basic we cannot do without it: a claim is (or is to be considered) true or false. What further simple or clear properties of claims may be of use? One good candidate is their linguistic structure.

Some consider linguistic structure enough. They set out formal systems of reasoning—logics—solely in terms of linguistic form. That is, they give syntactic forms of claims that are deemed true for every possibility and syntactic forms of inferences that are deemed valid. These are then taken as the norms of reasoning. This, they say, is a way to extricate ourselves from the possibility-inference circle.

Though it may allow for a way out of the circle, it promises no insight or further clarity about inferences or possibility. It leaves the circle only to embrace intuitive justifications for why we should accept claims of those particular forms as always true and inferences of those forms as valid. We are back at accepting an informal understanding of valid inference or an informal notion of possibility as the basis of our reasoning. This is not progress in understanding.

If, then, we look at the structure of claims, it is to reduce possibilities and inferences to simpler semantic notions.

Propositional logics

We begin by looking at very gross structures of claims: how we can build claims from other claims.

It is traditional to consider four ways to form new claims from other claims using the words, or as they are called *connectives*, "and," "or," "not," and "if . . . then . . .". For example, from "Ralph is a dog" and "George is a duck," we can form "Ralph is a dog and George is a duck," "Ralph is a dog or George is a duck," "Ralph is not a dog," "If Ralph is a dog, then George is a duck."

These connectives allow for a fairly rich analysis of reasoning. Once we have set out the semantic assumptions we will employ, we can ask whether other words that we apparently use to form new claims from old, such as "but," "despite," "nonetheless," "therefore," "in as much as," and "because," can be assimilated to one or a composition of these original four. Doing that we will have a better idea of the scope of our analysis of inference and possibility.

Since we intend to state explicitly the semantic assumptions about how these connectives work, we replace the ordinary words with symbols such as "∧ " for "and," "∨ " for "or," "¬ " for "not" as in "it is not the case that," and "→ " for "if . . . then . . .". These are the *formal connectives*. We can then be clear about what we mean when we say that we can make up new claims from others using these by defining a formal language that will give us the framework, the skeleton of all claims that we can consider in this analysis of reasoning.

We take the letters p_0, p_1, . . . to stand for claims; they are the *propositional variables*. Each is meant to stand for a claim that we consider to be, relative to our semantic assumptions, an unanalyzable whole; we call these the *atomic claims* or *atomic propositions*. We also need to be able to refer to these variables and the formulas we build up from them, so we employ *metavariables* A, B, C, ... Then we define a formal language $L(¬, → , ∧ , ∨ , p_0, p_1, . . .)$ by taking the analogue of a sentence in English to be a *well-formed formula* or *wff*:

a. Each of (p_0), (p_1), (p_2), . . . is a wff , an *atomic wff* and has *length* 1.

b. If A, B are wffs of length n, then each of $(¬A)$, $(A→ B)$, $(A∧ B)$, $(A∨ B)$ is a wff of length $n+1$.

c. A concatenation of symbols is a wff iff it is a wff of length n for some n.

The simplest analysis of the connectives we can make is to ignore all aspects of claims except truth-value and form, evaluating compound claims by:

A∧ B is true if and only if both A and B are true.

A∨ B is true if and only if A or B or both are true.

¬ A is true if and only if A is false.

A→ B is true if and only if A is false or B is true.

These conditions are usually set out in tabular form, and are called the *classical truth-tables* for the connectives:

A	B	A ∧ B
T	T	T
T	F	F
F	T	F
F	F	F

A	B	A ∨ B
T	T	T
T	F	T
F	T	T
F	F	F

A	¬ A
T	F
T	F
F	T
F	T

A	B	A → B
T	T	T
T	F	F
F	T	T
F	F	T

Schematically, we have:

$$L(\neg, \rightarrow, \wedge, \vee, p_0, p_1, \ldots)$$

↓   *realization*

$$\{\text{real}(p_0), \text{real}(p_1), \ldots, \text{claims formed from these using } \neg, \rightarrow, \wedge, \vee \}$$

↓   *valuation +*
        inductive definition of truth

$$\{ \text{true}, \text{false} \}$$

The first step, assigning claims to (some of the) variables, is called a *realization*. The *semi-formal language* is the collection of complex claims formed from the atomic ones by the rules for defining wffs. The second step, called the *valuation*, is an assignment of truth-values to the atomic claims, from which the truth-value of every semi-formal wff follows inductively by the classical truth-tables. The way in which we assign truth-values to the atomic claims cannot be taken into account in our analysis because it is not for the logician to say which atomic claims are true and which are false. The whole is a *model* of *classical propositional logic*.

Example 3 Ralph is barking or Juney is barking, yet Ralph is not a dog.

Howie is a cat or Howie is afraid of Juney.

If Howie is afraid of Juney and Ralph is not a dog,
    then Juney is a dog.

So Juney is a dog.

Analysis To use classical propositional logic to evaluate this inference, we first need to agree that "Ralph is barking," "Juney is barking," "Ralph is a dog," "Howie is a cat," "Howie is afraid of Juney," and "Juney is a dog" are claims, indeed atomic claims. We also need to agree that it is appropriate to formalize here both "and" and "yet" as $\wedge$, "not" as $\neg$, and "if . . . then . . ." as $\rightarrow$. In this example that seems O.K., but this first step is not always so straightforward.

So let's agree that the example is suitable to formalize as the following semi-formal wffs:

Premises: (Ralph is barking $\vee$ Juney is barking) $\wedge$ $\neg$(Ralph is a dog)

Howie is a cat $\vee$ Howie is afraid of Juney

(Howie is afraid of Juney $\wedge$ $\neg$(Ralph is a dog))
$\rightarrow$ Juney is a dog
Conclusion: Juney is a dog.

Under one choice of assigning claims to propositional variables, this is a realization of:

Premises: $(((p_0) \wedge (p_1)) \vee (\neg(p_2)))$

$((p_{47}) \vee (p_{312}))$

$(((p_{312}) \wedge (\neg(p_2))) \rightarrow (p_{4318}))$

Conclusion: p_{4317}

We need not realize every propositional variable; we consider only the claims in the reasoning we are analyzing.

We don't know which, if any, of these atomic claims are true. But we can analyze whether the inference is valid by looking at all the ways the premises could be true. For instance, we can suppose that the atomic claims "Ralph is barking," "Juney is barking," and "Howie is afraid of Juney" are true, and "Howie is a cat" and "Ralph is a dog" are false, so the first and second premises are true. Since those assignments make the antecedent of the third premise true, the consequent of the third premise must be true, if the third premise is true. Hence, for that assignment of truth-values, "Juney is a dog" is true. But if we suppose instead that "Ralph is barking," "Juney is barking," and "Howie is afraid of Juney" are true, while "Ralph is a dog" and "Howie is a cat" are both false, then the first and second premises are true; taking "Juney is a dog" to be false, the third premise is true, too. But that makes the conclusion false. So we have a model in which the premises are true and conclusion false; so the inference is not valid.

Formalizing reasoning in classical propositional logic, a possibility is reduced to a way to assign truth-values to atomic claims, resulting in a model for the logic. To say that an inference is valid relative to classical propositional logic is to say that there is no model in which the premises are (taken to be) true and the conclusion false. That is, there is no way to assign truth-values to the atomic claims in the semi-formal language that will make the premises true and the conclusion false. We have a reduction of the notion of possibility to simpler notions.

Still, this analysis invokes our informal notion of possibility, for we talk about any possible assignment of truth-values. But this is a restricted and much simpler notion than the idea of possibility in general. In an inference, we have a finite number of premises plus a conclusion, and only a finite number of claims appear in those, say n. There are 2^n different ways to assign truth-values to those, which we can inspect in a tabular form when n is small enough. When n is very large or where we choose to work with infinitely many premises, as we sometimes do in mathematics, we must rely on the method of specification of the atomic claims in them to guide us in how we can assign truth-values to them. The reduction of the notion of validity to simpler notions will then be as clear and convincing as our understanding of large numbers or an infinite number of premises.

We can expand our analysis to take into consideration further aspects of claims, such as how we may come to know them, or their subject matter, or how likely we deem them to be true. We do that by making structural decisions about the nature of the additional aspect of claims we are considering and then factoring those into slightly more complicated truth-tables. For each such analysis, that is, for each such logic, there is a corresponding notion of a model, which formalizes the notion of possibility and, hence, a notion of valid inference.[4] Depending on the semantic notions involved, the reduction of the notions of possibility will be more or less clear and helpful.

We can determine the forms of inferences that are valid for all such propositional logics. Perhaps these constitute the general form of valid inferences *relative to these connectives* that any reasoning creature would have to employ, as I discuss in "Why Are There So Many Logics?" in this volume. But it is not obvious that this constitutes an analysis of all reasoning even with claims as wholes, for there is no reason to think that for every connective of claims there is such a logic that aptly formalizes it.

Many logicians go further in abstracting from ordinary reasoning. They eliminate the realization part of models and say that a model is simply a valuation. In many textbooks a model for classical propositional logic is said to be an assignment of truth-values to the propositional variables. That is clearly wrong: a variable cannot be true or false. But the abstraction is a harmless and useful way to investigate the schematic form of valid inferences, for given an assignment of truth-values to the variables, we can find claims for the variables to which we assign those values. The forms of inferences that are valid, that is the formal language correlates of inferences any of whose instances (realizations) will be valid, can be determined more easily by considering such abstracted versions of models.

We now have analyses of possibility and valid inference for reasoning in the restricted context of paying attention to only claims as wholes and ways to build claims from them in a regular way using certain connectives. Each analysis, that is, each logic, reduces the notion of possibility to what are claimed to be simpler notions, which includes the truth or falsity of atomic claims. But we are far from a sufficient analysis of possibility. For example, consider:

All men are mortal. Socrates is a man. Socrates is not mortal.

All dogs are brown. Some dogs are green.

Snow is white. Anything that's white isn't black. Snow is black.

We know that each of these collections of claims is not a description of a way the world could be because we know—informally—that each is or leads to a contradiction. But we cannot show that any of them entails a contradiction if we are restricted to considering the forms of the claims relative to only propositional connectives. We need a way to understand possibilities that takes into account the internal structure of what we have so far taken to be atomic claims.

Predicate logic

There are several ways people have chosen to parse the internal structure of claims in analyses of reasoning. Each depends on a particular view of the world. Let's look here at the way that has been dominant in logical studies during the last one hundred years.

Consider the claim "Spot is a dog." It has no structure relative to the propositional connectives. But we can parse it as we learned in school into a name and predicate: "Spot" and "is a dog." Extending the notion of predicate, we can say that "Spot loves Dick" can be parsed as two names "Spot" and "Dick" and a predicate "loves."

Modern predicate logic takes this way of parsing claims as fundamental. A *predicate* is a claim with the names (and pronouns) deleted, with clear marking for where those came from and where those or other names could be inserted.[5] Thus, "Spot is standing between Zoe and Dick" yields the predicate "— is standing between — and —".

This way of parsing atomic claims is based on the view that the world is made up of things, individual things. Names and pronouns stand for things, and predicates are what are said to true of things. Things are said to *satisfy* a predicate, or a predicate is said to *apply to* certain things, if when names of those things are put into the blanks of the predicate, a true claim results.

This method of parsing claims facilitates evaluating reasoning with assertions about all or some things satisfying a predicate. The symbols "$\forall$" and "$\exists$" are used as formal counterparts to "all" and "exists" (or "some"). For example, "Everything loves Ralph" is formalized as "$\forall x\,(x$ loves Ralph)", and "There is something that is bigger than Ralph" is formalized as "$\exists x\,(x$ is bigger than Ralph)", where the variable x is used to allow us to refer to things in a general way.

Here, too, we set out the structure of claims we'll consider in this logic by defining a formal language. We start with *predicate symbols* $P_0^1, P_0^2, \ldots, P_1^1, P_1^2, \ldots$ which stand for predicates, where the superscript indicates how many blanks are in the predicate that have to be filled with names or variables. We take *name symbols* $c_0, c_1, \ldots$ to stand for names and use $x_0, x_1, \ldots$ as *individual variables*. The name symbols and variables together are called *terms*; we use the metavariables $t_0, t_1, \ldots$ to stand for any of those. The formal language also has the *quantifiers* "$\forall$" and "$\exists$", parentheses, and the connectives from our propositional logic. We define the *well-formed-formulas* or *wffs* :

    a. For each $i \geq 0$, and $n \geq 1$, $(P_i^n(t_1, \ldots, t_n))$ is a wff, an *atomic* wff, which has length 1.

    b. If A, B are wffs of length n, then each of $(\neg A), (A \to B), (A \wedge B), (A \vee B)$ is a wff of length $n+1$.

    c. If A is a wff, then for any $i \geq 0$, each of $(\forall x_i\, A)$ and $(\exists x_i\, A)$ is a wff of length $n+1$.

    d. A concatenation of symbols is a wff iff it is a wff of length n for some n.

Unlike propositional logic, not every wff of the formal language can correspond to a claim. For example, $P_0^2(x_0, x_1)$ does not have the structure of a claim, because replacing the predicate symbol with, say, "is bigger than," we get "x_0 is bigger than x_1"; that is not a claim until we say what things x_0 and x_1 stand for. Only formulas in which every variable in the formula is quantified correspond to claims; these are the *closed wffs*. Thus, $\forall x_0 \exists x_1 \, P_0^2(x_0, x_1)$ can correspond to the claim "For everything there is something that is bigger than it." And "$\exists x_1 \, P_0^2(x_0, x_1)$"has the structure of the claim "There is something bigger than Spot." And $P_0^2(c_0, c_1)$, which has no variables, can correspond to "Dick is bigger than Spot."

A *realization* is an assignment of linguistic predicates to some or all of the predicate symbols and names to some or all of the name symbols. The result of replacing the predicate symbols with predicates and name symbols with names in a wff is a *semi-formal wff*, a formalized fragment of English or some other ordinary language.

To give semantics for the wffs of the semi-formal language wffs, we first take some collection of individual things to be what we're reasoning about; we call that the *universe* of the realization For the generality, we would prefer to take the universe be all things, but there are good reasons why we don't.[6] Then we assume that each name, such as "Spot" or "Dick," stands for one of those things. Then every atomic semi-formal wff can be understood as a claim, true or false, when a reference is supplied for each variable in it; this is what we call a *predication*. Thus, "x_1 is bigger than Spot" is an atomic claim when we say what "x_1" is to refer to Tom. These are the minimal semantic assumptions we need. We don't say "what a predicate is" or "what a predicate means" beyond agreeing on what things it is true or false of, leaving our logic open to many such interpretations.

Then, as for propositional logic, we have a method to extend the semantic analysis to compound propositions by using the tables for the connectives along with a method of evaluating the quantifiers inductively treating "$\forall x$" as meaning that for every reference that we could supply for x the result is true, and "$\exists x$" as meaning that for some reference that we could supply for x the result is true.

Schematically, a *model of classical predicate logic* is:

$$L(\neg, \rightarrow, \wedge, \vee, \forall, \exists, x_0, x_1, \ldots, P_0^1, P_0^2, \ldots,$$
$$P_1^1, P_1^2, \ldots, c_0, c_1, \ldots)$$

$\downarrow$ *realization*

$$L(\neg, \rightarrow, \wedge, \vee, \forall, \exists, x_0, x_1, \ldots;$$
realizations of predicate and name symbols)

universe: specified in some manner

$\downarrow$ *assignments of references as needed*;
*assignments of truth-values to atomic
predications*; *truth-tables for* $\neg, \rightarrow$,
$\wedge, \vee$; *evaluation of the quantifiers*

$\{\mathsf{T}, \mathsf{F}\}$

For example, we might have:

$$L(\neg, \rightarrow, \wedge, \vee, \forall, \exists, x_0, x_1, \ldots, P_0^1, P_0^2, P_0^3, \ldots, P_1^1, P_1^2, P_1^3,$$
$$\ldots, c_0, c_1, \ldots)$$

$\downarrow$

$$L(\neg, \rightarrow, \wedge, \vee, \forall, \exists, x_0, x_1, \ldots; \quad -\text{is a dog}, -\text{is a cat},$$
$-$eats grass, $-$is a wombat, $-$is the father of$-$; Ralph,
Dusty, Howie, Juney)

universe: all animals, living or toy

$\downarrow$

$\{\mathsf{T}, \mathsf{F}\}$

In the case of predicate logics built on other propositional logics, additional aspects of predicates and names are incorporated into the semantics, such as the ways we can come to know references or the subject matter of predicates and names. These are sometimes called the *content*, or *connotation*, or *sense* of the names and predicates. To simplify this discussion I'll discuss only classical predicate logic.[7]

Example 4   Juney is a dog
     Juney likes Richard L. Epstein.
     So all dogs like Richard L. Epstein.
     Analysis   Informally, we recognize that this inference is not valid. It might be that Juney is a dog, Juney likes Richard L. Epstein, but there is some other dog, some unnamed mongrel who has never learned to love, that doesn't like me.

For our analysis in classical predicate logic, we first need to assume that each of the sentences in the inference is a claim. Further, we need to agree that "— is a dog" and "— likes —" are predicates, and that "Juney" and "Richard L. Epstein" are names, and that "all" can be understood as "$\forall$" here. We also have to have some way to formalize the conclusion, since "dogs" is not a name and there isn't any variable there. It's not obvious and requires some motivation to see that with the semantic assumptions of this logic we can formalize the conclusion as "$\forall x$ (x is a dog $\rightarrow$ x likes Richard L. Epstein)". So the semi-formal version of the inference is:

Premises   Juney is a dog
             Juney likes Richard L. Epstein

Conclusion $\forall x_1$ (x_1 is a dog $\rightarrow$ x_1 likes Richard L. Epstein)

Under one choice of assignment of predicates to predicate symbols and names to name symbols, this is a realization of:

Premises $P_0^1 (c_0)$

             $P_{14}^2 (c_0, c_1)$

Conclusion $\forall x_1 (P_0^1 (x_1) \rightarrow P_{14}^2 (x_1, c_1))$

We need not realize every predicate symbol and every name symbol; we consider in our realization only all the parts of the inference we are analyzing.

Now our analysis proceeds as the informal one, though we have a choice for what universe we'll use: all animals, living or dead; all dogs, living or dead; all living dogs; all living dogs in North America; all animals living or dead; . . . . Let's suppose we take the last. Then to show that the inference is invalid, consider a model in which "Juney" names some object, "Richard L. Epstein" names some object, and the predications "Juney is a dog" and "Juney likes Richard L. Epstein" are true, and further there is an object in the universe such that when x_1 is taken to refer to it the predication "x_1 is a dog" is true and the predication "x_1 likes Richard L. Epstein" is false. In that case, the premises are true and conclusion false. So our formal logic also classifies this inference as invalid

A possibility in predicate logic is a way to assign truth-values to the atomic claims in a realization. But the atomic claims here include wffs such as "x_1 is a dog" when a reference is supplied for "x_1". So not only truth-values are assumed, but also some idea of naming or picking out references from the universe. These two semantic notions plus the evaluation of the connectives and the quantifiers are all that we need to understand possibilities, once we have agreed on the formalization of any particular inference.

As the last example shows, this understanding of possibilities correlates with how we informally analyze inferences that can be formalized in this logic. Here, as for propositional logic, any choice of assignments of truth-values to atomic predications must be allowed in order to have a rich enough notion of possibility in this logic. We must allow a model in which "Marilyn Monroe was a man" and "Napoleon lived in California" are taken as true. This is not to say they are true; rather, a possibility is a model, a description (in terms of the linguistic elements we recognize) of a way the world could be, and that is reduced to an assignment of truth-values to atomic predications. We need only consider the predicates and names in the inferences; to see that the semi-formal version of Example 4 is invalid, we don't need to consider whether "Marilyn Monroe is a man" is true or false. Descriptions are only as full as needed in the analysis of the inference at hand.[8]

Example 5 All men are mortal.
  Socrates is a man.
  Therefore, Socrates is mortal.

Analysis We informally recognize this as valid. But if asked to explain why, we are at a loss to say more than that's just what the words mean—how could all men be mortal and Socrates not be mortal if he is a man?

Classical predicate logic can make this clearer. We can formalize the example:

Premises: $\forall x_1$ (x_1 is a man $\rightarrow$ x_1 is mortal)
              Socrates is a man.

Conclusion: Socrates is mortal.

Any object in the universe to which "—is a man" applies must also satisfy "—is mortal". There must be an object in the model to which "Socrates" refers, and "—is a man" applies to it because of the second premise. So that object satisfies "—is mortal". So there's no way for the premises to be true and conclusion false.

Does classical predicate logic really give us a reduction of the notion of valid inference? To show that an inference is valid we have to consider all possible models of a particular semi-formal language. Often, as in the last example, this can be done by considering general properties of the models. However, we often reason without specifying each and every object in the universe, as when we reason about all the pigs in Denmark. We rely on our notion of naming: given that pig, we could name her "x_1" and then "x_1 is black" is true or false. We can claim that the reduction of possibility and valid inference here is simpler and clearer than the full informal notion in any particular case only to the extent that we believe our notion of naming is simpler and clearer. We have made some progress. In terms of the semantic and syntactic assumptions of this logic, what is logically possible and what is a valid inference have been reduced to just the truth-values of atomic claims and to our notion of naming, along with our evaluation of the connectives and quantifiers.

But it is only progress and hardly the whole story of logical possibilities.

Example 6 Ralph is a bachelor. So Ralph is married.

Analysis We know that these two claims can't describe a way the world could be because we know that "bachelor" means in part "not married." But "—is a bachelor" and "—is married" are distinct predicates that have no structure we recognize in predicate logic, so we could have a model in which there is an object named by "Ralph" and both predicates "—is a bachelor" and "—is married" apply to it. This analysis shows that we should not take both "—is a bachelor" and "—is married" as atomic predicates in a model if we wish to respect our informal notion of possibility.

Alternatively, we can incorporate our assumption about the meaning of those predicates into our analysis by requiring that the semi-formal claim "$\forall x_1$ (x_1 is a bachelor $\rightarrow \neg($ x_1 is married$)$)" be true in any model in which we use these predicates. Such a definition is not part of an inference, for a definition is not true or false, but only apt or inept, good or bad. It is an assumption that underlies the choice of how we will formalize.

Example 7 All dogs are brown and some dogs are green.

Analysis We know that it's not possible for all dogs to be brown and some dogs to be green, understanding these predicates as "almost entirely brown" and "almost entirely green." But "—is brown" and

"—is green" have no internal structure recognized in predicate logic, so they have to be taken as atomic predicates. Hence, we could have a model in which "—is a dog" and "—is green" and "—is brown" apply to every object. Here we don't want to say that we can't use both "—is green" and "—is brown" together in our semi-formal languages, for we cannot derive "—is green" from "—is brown". Nor are we lacking a definition. Rather, it is our understanding of the use of these words relative to our experience of the world that convinces us that the example is not a possibility. We can formalize part of that experience with two claims:

$\forall x_1 (x_1 \text{ is brown} \rightarrow \neg (x_1 \text{ is green}))$

$\forall x_1 (x_1 \text{ is green} \rightarrow \neg (x_1 \text{ is brown}))$

If we restrict the analysis of inferences in which these predicates appear to those models in which these claims are true, then the example will not be classified as a possibility.

These last two examples show that we can adapt our formal logic to analyze possibilities and validity in cases where the semantic and syntactic resources of predicate logic are apparently too limited. The next example, however, shows a real limitation of predicate logic in analyzing possibilities.

Example 8   Snow is white.
Anything that's white isn't black.
Snow is black.

Analysis   We cannot use classical predicate logic to show that these three claims comprise a contradictory description of the world. Snow, as used in the first claim, is not an individual thing; it cannot be in a universe of predicate logic; it cannot serve as reference for a variable. It is a mass, and the assumption that the world is made up of individual things does not take into account the world being made up in part of masses that are not things. Hence, we can't take "— is snow" as a predicate, for no name of a thing could fill the blank to make that true. A different logic, Aristotelian logic, can be used to show that this example is not possible, for that logic is based on the view that the world is made up in part of masses. But no way of melding that logic and any predicate logic has been devised.[9]

When we consider the internal structure of atomic claims, our analyses of reasoning will depend on some view of the nature of the world. So our formalization of reasoning will accommodate only reasoning compatible with that metaphysics. At best we can hope to combine various formalizations that depend on different but compatible metaphysics into one logic, such as a logic that is based on seeing the world as made up of things and of masses.

Example 9   Birta is a dog.
Bon Bon is a donkey.
Therefore, Birta did not give birth to Bon Bon.

Analysis   To formalize this inference in classical predicate logic we replace "not" with the connective "‐|", we take "Birta" and "Bon Bon" as names, and we take "—is a dog," "—is a donkey," "—gives birth to—" as atomic predicates. Since these predicates have no structure that we recognize in predicate logic, we can have a model in which any truth-value can be assigned to any predication involving them. In particular, we could have a model in which the premises of this example are true and conclusion false, so the inference is not valid.

That's just to say our logic recognizes that a dog giving birth to a donkey is a possibility. Is that right? There seems to be nothing in the meaning of these predicates that determines whether it is possible for a dog to give birth to a donkey. The predicates are atomic, and anyone who wishes to show that a dog giving birth to a donkey is not a logical possibility must appeal to some other standard of reasoning to show that.

Conclusion

We have made some progress in understanding possibilities and valid inferences. We look at an inference and try to agree on what semantic assumptions we use in informally analyzing it. Then we ask whether we have or can devise a formal logic based on those semantic assumptions. If yes, then we try to formalize the inference in that logic relative to those semantic assumptions. If we are successful, we have clarified in this instance our claim that we do or do not have a logical possibility and whether we do or do not have a valid inference.

In some cases it is quite easy to formalize an inference. In other cases it is quite difficult, requiring us to question our understanding of language and the world. And in other cases there is no formal logic we can use; we have only our informal notions on which to rely.

To judge an inference as valid or invalid requires us to be explicit about the semantic assumptions we are making so that we can be clear enough to reason well together. When we can agree, we have made some progress.

Appendix 1 Using logical possibilities in philosophy

Twin Earth

Hilary Putnam describes what he calls "Twin Earth" in an analysis of the nature of necessity. Twin Earth is very much like earth except that a substance there has the usual properties we ascribe to water but isn't H_2O. It is, he says, a possibility which has to be considered in theories of necessity.[10]

Every time I read Putnam's description, I find it too bizarre. I can't conceive of such a world. How could Putnam convince me otherwise? [11]

Conceivability isn't the issue. Perhaps you can't conceive of a dog giving birth to a donkey, but it is a possibility we have to take into account in some of our reasoning *if* we can show that no contradiction follows from it, which we have done relative to the assumptions of classical predicate logic.[12]

Similarly, when challenged, Putnam needs to show that no contradiction follows from his description of Twin Earth. But his job is much harder, for he cannot use the resources of predicate logic. "Water" stands for a mass, and predicate logic is not suitable to formalize reasoning about masses. Aristotelian logic is suitable, but much of the rest of Putnam's description requires the resources of predicate logic. And we have no accepted formal logic that combines Aristotelian logic with predicate logic.

To the challenge that his description is not a logical possibility, Putnam can reply only by showing informally that it leads to no contradiction. That is not easy, and he has not done it. It might require assumptions about inferences and possibility that are not compatible with his views in that article. We do not know.[13]

Nothing

Jared Diamond relates how as a student at Harvard his professor Paul Tillich stymied him and his classmates with the question "Why is there something, when there could have been nothing?", using that question to lead to belief in the existence of a creator of all.[14]

But the question supposes what needs to be shown: There could have been nothing. All of our experience seems to deny that. Yes, there might not have been Socrates, or atoms, or anything we currently know. But nothing?

A colleague of mine took me to task for saying this. After all, he said, it's logically possible that there could have been nothing, and to deny that possibility is to offer one answer to the question.

But how do we know that's logically possible? It's not a matter of taking predicates we use in our daily lives and assigning different truth-values to predications using them. Predicate logic is no guide to reasoning about nothing.[15] No formal logic, no reduction of the notion of possibility and inference, is available to guide us in reasoning about nothing. And informal methods for doing that seem as suspect as assuming that there could be nothing.[16]

The burden of proof in establishing possibilities

Who has the burden of proof to establish that a collection of claims describes a way the world could be? If we can invoke a formal logic to analyze a collection of claims, we can justify that a particular collection is consistent. If we have no such logic, then the burden of proof is on the person claiming that the collection of claims is consistent. Someone might say that putting the burden of proof on the person who asserts the existence of a possibility would seriously impede the use of thought experiments. But if it's bizarre, we are owed a justification. If you say you saw a pig fly out of the sky last night and hover three feet above you for two minutes, I'd ask for some evidence. If you say that Putnam's description of Twin Earth is consistent, I'd ask for some evidence.

D. H. Mellor in "Theoretically Structured Time" also believes that the burden of proof is on the person who claims that a description is possible:

> I have in any case a general objection to this fashionable kind of argument by fantasy. It presumes to show something possible by describing an imaginary world in which we should apparently be inclined to believe the possibility actual. That may serve to settle verificationist qualms about the sense of supposing (e.g.) that the Universe has doubled in size overnight. But to show possibility as well as sense, one has also to show the imaginary world possible, which a merely plausible sketch of it does not do. Impossibilities are all too easy to make plausible, as mathematics shows: '$10^{23} + 1$ is prime', for example, is a plausible enough statement; but it may be false nonetheless, and if it is, it is impossible. But even if Newton-Smith's fantastic worlds were possible, the plausibility in them of multiple- or closed-time hypotheses would not show that they in turn might be true. For only if they too are really possible would their *prima facie* plausibility in those worlds give reason to think them true. In the face of contrary arguments *a priori*, such as the one I have alluded to against causal loops, fantasies supply no evidence for possibility at all.
>
> Nor can fantasy be propped up in this respect by physics. Certainly the fashion for it is reinforced by that for taking current physics to be at once the arbiter of truth and subject to a neurotic post-positivist conviction that it will all end up falsified and replaced by something else. This daft combination of attitudes undoubtedly encourages unrestrained fantasizing about conceivable future physics, whose cognitive authority is then invoked to establish the present possibility of its consequences, however far-fetched. But these are quite obviously worthless trains of thought. Unless our acceptance of

current physics commits us to its truth, it has no cognitive authority over us at all, and a merely conceivable future physics cannot have more authority over us than our own current physics has. In any case, even acceptance of physics as true does not preclude a mistaken acceptance of the impossible. No doubt we *ought* only to accept the possible as true; but then rational acceptance will have to wait on proof of possibility, not supply it.   pp. 66-67

But W. H. Newton-Smith, in "Reply to Dr. Mellor," disagrees:

To show possibility, he says, 'one has to show the imaginary world possible, which a merely plausible sketch of it does not do.' Mellor has misunderstood the structure and point of such arguments. One deploys them in relation to a contentious idea (e.g. bringing about the past, branching time) which has not been shown by a knock-down argument to be incoherent. The first stage involves giving a description which it is uncontentious to claim characterizes a possible world. . . . The second stage is to consider what further explanatory description might be made of the possible world. . . . The third stage is to argue that in the possible world the evidence in fact supports invoking the contentious idea . . . *If* anyone had a proof of the incoherence of the contentious idea it could be deployed here to defeat the argument. The point of such arguments is twofold. First, they still doubts of a verificationist sort concerning the contentious idea. Secondly, they are intended to *persuade* one who feels that there is some absurdity in the contentious idea that this is not so. As Dummett points out in the prelude to his own argument by fantasy in the paper favourably cited by Mellor ["Bringing about the Past"], if the idea of bringing about the past is logically absurd, that absurdity should show up no matter how things turn out. A description of how things might turn out as in my stories or Dummett's enables us to see that an idea which in the abstract seems absurd could in fact have fruitful application and is not absurd.

Deploying fantasy physics in this context does not involve assuming the worthless train of thought outlined by Mellor. The structure of the argument is as characterized above. One describes a way things could turn out to be . . . and one shows that in such a context it would be reasonable to posit the applicability of a contentious idea . . . Certainly Mellor is correct in claiming that even acceptance of present-day physics as true does not preclude a mistaken acceptance of the impossible. However, it remains true that the success of modern physics gives us a reason to think that it is not based on impossible notions.   pp. 69–70

Compare also what Jonathan Bennett says in *A Study of Spinoza's Ethics*:

> I would go further and argue positively that it is unreasonable to
> believe that there must be a maximally excellent being. Plantinga
> notes that if his thesis is possible, then it is true [Alvin Platinga, *The
> Nature of Necessity*, p. 219f] (In the most widely used systems of
> modal logic, any proposition of the form 'Necessarily P' is true if it is
> possible.) And this contributes to his sense of entitlement to believe
> it. Why, after all, should he give up his view that it *could* be true
> that there must be a maximally excellent being? Leibniz made a
> similar point about the premiss which he added to Descartes'
> ontological argument, namely, that there could be a being which
> fits Descartes' definition: 'One is entitled to presume the possibility
> of any Being . . . until someone proves the contrary.'
>
> That principle about the onus of proof looks right. But properly
> used it counts against Leibniz and Plantinga, not for them. Someone
> who says it is possible that there must be an F being is basically
> asserting that there must be an F being, and is thus asserting an
> infinity of denials of possibility: of every world description, which
> excludes the existence of an F being, he is saying that it is impossible
> for there to be such a world as that. So if F is maximal excellence, he
> will say that there could not possibly be a world in which the only
> concrete objects were time and space and portions of matter. I reply
> that I am entitled to presume the possibility of such a world until
> someone proves the contrary, and I add that it is not reasonable to
> believe in the impossibility of such a world without having positive
> reasons for doing so.[17] pp. 71–72

Bennett, Platinga, Leibniz all hold that for a collection of claims to be a
description of a way the world could be it must be consistent. They all take it
that the burden of proof is to show that the collection is not consistent. On the
face of it, they say, anything we postulate is possible, though Bennett restricts
that in his paper to "possibilities that do not themselves have modal concepts
nested within them."

Appendix 2 Logical consequence
Here I'll relate some of the recent debates about what's called "logical
consequence" to the analysis of valid inferences I've given.[18]

Form as a guide to valid inferences

William Hanson in "The Concept of Logical Consequence" writes:

> Other texts try to avoid modal notions and hence make use of truth
> and falsity simpliciter, rather than truth and falsity at possible worlds.

> They say that the conclusion of an argument is a logical consequence
> of the premises just in case no argument with the same logical form
> has true premises and a false conclusion. p. 366

To decide what is the logical form of a claim, we need to have in hand
the logical constants. But, as Hanson and others point out, there is no unique
collection of those that can be considered exhaustive. So Hanson offers his
criteria:

> In order to satisfy the requirements of generality and apriority, the
> semantic element is subject to two constraints on its selection of
> constants: (a) it must designate some terms as logical constants, and
> it will include among them some that appear in discourse on a wide
> variety of subjects (that is, what I earlier called ubiquitous terms); and
> (b) taken together, the terms it chooses must allow us to distinguish
> arguments exhibiting the resulting relation of logical consequence
> from those that do not exhibit it in a strictly a priori manner, to the
> extent that we can make such distinctions at all. The choice of logical
> constants is thus pragmatic in the sense that it is influenced by our
> goals that logic should exhibit the kinds of generality and apriority
> that I have discussed.

But there are no logical constants as he describes. English words, or
words of any other ordinary language, cannot be logical constants in this sense
because they are not used in a manner that is univocal. Consider the inference:

> Bring me an ice cream cone and I'll be happy. I'm not happy.
> Therefore, you didn't bring me an ice cream cone.

We recognize this as valid. And doing so we recognize that "and" here cannot
be formalized as the propositional logic "∧ ", for that would yield an invalid
inference. The word "and" should be formalized with "→ ".

We must abstract from our usual understanding of ordinary words if we
are to do anything like a formal analysis. Then it is the formal logical constants
that determine the form of a claim. But to formalize claims relative to these
constants, we must observe the most fundamental constraint on formalization,
which the example above illustrates:

> An informally valid inference must be formalized as a valid inference.

It is our understandings of possibilities and valid inference that determine
whether any particular formalization is good. The reliance on forms of claims
in this approach already assumes some informal notion of possibility.

In all the papers cited, logical form is understood as some predicate logic
form. But that's just to assume the hard work that needs to be done: why
should we accept the metaphysical assumptions needed for predicate logic
in order to formalize a particular inference?

The actual world suffices in place of possibilities

Some say we do not need to rely on possibilities in analyses of inferences. We need consider only the world as it is and vary our interpretations of the non-logical words in inferences. Consider, for example:

> Marilyn Monroe is a man.
> Every man has two feet.
> Therefore, Marilyn Monroe has a right foot.

The usual reasoning is that this is invalid: both "Marilyn Monroe is a man" and "Every man has two feet" could be true, while Marilyn Monroe could have two left feet, and that is easy to formalize in classical predicate logic. But rather than using possibilities, it's said we can see that this example is invalid by understanding "Marilyn Monroe" as referring to me, "is a man" as having its usual meaning, "has two feet" as meaning "has two dogs," and "has a right foot" as meaning "has a black dog"; the premises are true and the conclusion false in the actual world we live in, so the inference is not valid. We need not invoke possibilities.

We need to know the logical form of the inference before we can apply this method, for we need to know which terms we can vary the meaning of. As we saw above, that requires the use of some prior notion of possibility or valid inference.

Situations and possibilities

William Hanson considers the following example:

> $(\exists x)\,(\exists y)\,(x \neq y)$
> Therefore, $(\exists x)\,(\exists y)\,(\exists z)\,(x \neq y\ \&\ y \neq z\ \&\ x \neq z)$

To use classical predicate logic to show that this is invalid, we need that there is a model with just two things. Hanson worries that the platonist says that mathematical objects necessarily exist, so for a platonist there couldn't be just two things. To overcome this objection, he says we could argue:

> In determining logical truth and logical consequence we are concerned not only with truth values at all possible worlds but also with truth values at all subworlds of situations that appear as possible worlds. . . . For the sake of promoting generality, one is willing to accept as counterexamples to arguments not only full-fledged ways things might have been but also fragments of such ways. I believe that logicians often think of logical models in this way, and that doing so does not commit them to the view that each such model is itself a full-fledged possible world (that is, a way things might have been). For the mathematical platonist, doing so can be seen as introducing further generality into logic, a generality that comes from taking logic to be

applicable even to the tiniest and most bizarrely gerrymandered fragments of possible worlds.[19]

In all the examples above, in describing possibilities we considered only the words that appeared in the inferences. No one gives a description of a way the world could be that is fully complete, and no one could. One assumption we always make is that some of the words in the inference are clear enough and distinct enough in meaning from the other words in the inference to be taken as basic, allowing us to describe a possibility — or a part of a possibility if you like — by considering only claims in which those words appear.[21] In using predicate logic to analyze an inference we need consider only models in which we put just enough objects in the universe for every name in the inference to have a reference and whatever other objects are needed in order to do our analysis, as the examples above illustrate. If that does not satisfy the platonist, then the platonist justification of the inference as valid would be for any mathematician a *reductio ad absurdum* of platonism.

The same misconception of the use of possibilities in inference analysis appears in a view described by J. C. Beall and Greg Restall:[21]

> The world is made up of situations. They are simply *parts* of the world. Claims are true of not only the world as a whole, but some claims at least are true of situations. We will not spend time on the theory of situations and their individuation here: we will simply illustrate it. In the situation involving Greg's household as he writes this, it is *true* that Christine is reading the paper. It is also *true* that the stereo is playing. It is *false* that the television is on. It follows from this, and the fact that the television is in fact an inhabitant of the situation, that it is *true*, in this situation, that the television is off.
>
> Situations 'make' claims true and they 'make' others false. However, some situations, by virtue of being *restricted* parts of the world, may leave some claims undetermined. It is not true in this situation that JC is reading. It is also not false in this situation that JC is reading — that is, it is not true in this situation that JC is *not* reading. JC does not feature in this situation at all.
>
> It follows that the classical account of negation fails *for situations*.

A description of the world is never complete. We always look at some collection of claims that amount to our description: it's what we're paying attention to in our reasoning. We're reasoning about these claims. If someone says we should also consider some other claims, we'll do so, and then those, too, will be part of our description. A possibility is then just our description, so long as it is consistent and satisfies whatever other semantic conditions our logic imposes. Classical negation does not fail for claims just because they weren't originally under consideration.

Set theory as a reduction of the notion of possibility

Hilary Putnam says,

> The notion of possibility does not have to be taken as a *primitive*
> notion in science. We can, of course, define a structure to be
> *possible* (mathematically speaking) just in case a model exists
> for a certain theory, where the notion of a model is the standard set
> theoretic one. That is to say, we *can* take the existence of sets as
> basic and treat possibility as a derived notion.[22]

This is wrong, even for mathematical theories. It assumes that the
mathematical possibilities that are of interest are for a theory that has been
completely formalized, and very few mathematical theories have been. It also
assumes that there is a uniquely correct logic for which set theoretic models are
taken; presumably, given the era in which Putnam was writing and his other
writings suggest, first-order classical mathematical logic. And it also assumes
that there is a uniquely correct set theory. All these assumptions are wildly at
odds with mathematical practice.[23]

We could proceed as Putnam suggests to get some grasp of possibilities
relative to a set theory and logic, much as I have described for propositional
logic and for predicate logic. But it seems misguided to use set theory as a
foundation for the notion of possibility. The notion of the collection of all
subsets of a set seems no clearer than the notion of possibility.

Dedicated to Walter Carnielli

Notes

1. (p. 2) Some authors state the condition for validity as: a valid inference is one such that if the premises are true then the conclusion must be true; see, for example, J. L. Bell and M. Machover, *A Course in Mathematical Logic*, p. 5; Alfred Tarski, "On the Concept of Logical Consequence," p. 411; John Etchemendy, *The Concept of Logical Consequence*. This appears to be a careless way of making the definition given in this paper, for if it were taken literally then only a necessary claim could be the conclusion of a valid inference.

There is no reason to think that the notion of validity presented here is a common one or is abstracted from one that is more or less universally held by most people; compare the research in *"Truth" as Conceived by Those Who Are Not Professional Philosophers* by Arne Naess.

2. (p. 2) This is discussed in *The Fundamentals of Argument Analysis* in this series.

3. (p. 4) Some logicians have attempted to formulate how to reason when the information we have is or might be inconsistent. A few have argued that contradictions, such as there being square circles, are possible. But such a bizarre assumption is not needed for reasoning around contradictions, as I show in "Paraconsistent Logics with Simple Semantics," and it would leave us with no semantic basis from which to start our analysis of possibilities, as discussed in "Truth and Reasoning" in this volume.

4. (p. 9) This is spelled out in "A General Framework for Semantics for Propositional Logics" in this volume.

5. (p. 11) Those who take propositions to be abstract say that predicates are abstract, too, and that predicates as defined here only stand for or represent abstract predicates.

6. (p. 12) We have no way of reasoning about the collection of all things without contradictions, and we have no way to explain naming that is general enough to cover all things, as I discuss in *Predicate Logic*.

7. (p. 13) See *Predicate Logic* for the general framework for such predicate logics. See Stanisław Krajewski's and my "Relatedness Predicate Logic" for a development of one such logic.

8. (p. 15) This is justified by the Partial Interpretation Theorem in *Classical Mathematical Logic*.

9. (p. 17) See Chapter V.H of *Predicate Logic* for a discussion of why reasoning about masses cannot be formalized in a predicate logic. Or see Francis Jeffrey Pelletier and Lenhart K. Schubert, "Mass Expressions."

10. (p. 20) Putnam, "Meaning and Reference."

11. (p. 20) Paul Needham in "Microessentialism: What is the Argument?" says:

> What could Putnam's twin earth fantasy show beyond the triviality that sufficiently ignorant people might not be able to distinguish similar substances? If it assumes that two substances are distinct at the microlevel and yet share all their macroproperties, so that they can't be distinguished in terms of macroproperties, then the scenario is wildly implausible. Assuming that it is in some sense possible doesn't show it to be possible. p. 11

12. (p. 20) Some logicians do think that what is possible is what is conceivable:

> It is thus in principle inconceivable—that is, it is literally impossible —that the same events should occur at two distinct moments of time.
> Nicholas Rescher and Alasdair Urquhart, *Temporal Logic*, p. 151

13. (p. 20) One colleague commented to me:

> The Chinese recognized a kind of stone, namely, jade. After chemical analysis was developed, it turned out that some jade was nephrite and some was a different mineral, jadeite. So, we know what it is like to regard something as a natural kind that turns out to have samples with different microstructures. The property of being water is the property of being H_2O. But the property of being watery is the property of being transparent, tasteless, having low viscosity, etc. Once these surface properties are distinguished from the microstructural properties, we can entertain the possibility that not everything watery is water, just as we can understand that not all jade is nephrite.

But using an argument by analogy to show that a description is consistent is not adequate, for it we would need to show that the differences between the two sides of the comparison do not matter.

14. (p. 20) Diamond, "The Religious Success Story."

15. (p. 20) There are two uses of "all" in English: with or without existential import. If we take "∀" to have no existential import, as is usually the case in formalizing, then every claim beginning with "∀" is true about nothing, and every claim beginning with "∃" is false. If we take "∀" to have existential import, then in reasoning about nothing every claim beginning with a quantifier is false. In either case the resulting logic is trivial and no help in reasoning about nothing. This is just an indication that the use of variables does not make sense when applied to a model with empty universe, for there is no story of naming and predication for that. (Andrzej Mostowski, "On the Rules of Proof

in the Pure Functional Calculus of the First-Order," axiomatizes the first reading invoking set-theoretic interpretations of predicates, but that does not explain how he understands the use of variables compatible with how they are used in models with non-empty universes.)

A platonist must interpret the original question as asking why there are concrete things, since he or she believes that abstract objects such as numbers necessarily exist. In that case, the issue is not how to reason about nothing, but why we should believe that there are abstract things.

16. (p. 20) For an example of confusions that arise in trying to reason about nothing, see Robert Nozick, "Why Is There Something rather than Nothing?"

17. (p. 23) Bennett's parenthetical comment about systems of modal logic is wrong: it only holds for S5 if we take it as $\Diamond\Box A \supset \Box A$. It is correct for all extensions of T (reflexive frames) if we take it as $\Diamond\Box A \vdash \Box A$, but that is a strange reading.

18. (p. 23) See, for example, John Etchemendy, *The Concept of Logical Consequence*; William Hanson, "The Concept of Logical Consequence"; Timothy Bays, "On Tarski on Models"; José M. Sagüillo, "Logical Consequence Revisited"; Greg Ray, "Logical Consequence: A Defense of Tarski"; G.V. Sher, "Did Tarski Commit "Tarski's Fallacy"?"; Stephen Read, "Formal and Material Consequence."

19. (p. 26) Hanson, "The Concept of Logical Consequence," p. 388.

20. (p. 26) See the theory of formalizing presented in *Predicate Logic* and the Partial Interpretation Theorem in *Classical Mathematical Logic*.

21. (p. 26) Beall and Restall, "Logical Pluralism," pp. 475–493.

22. (p. 27) Putnam, "What is Mathematical Truth?", p. 71.

23. (p. 27) See my *Classical Mathematical Logic* and "Mathematics as the Art of Abstraction."

A General Framework for
Semantics for Propositional Logics

We cannot pay attention to all aspects of propositions in our reasoning. Our choice of which aspects we consider significant determines our notion of truth for compound claims and hence our logic. We can devise a general framework for semantics for propositional logics based on how we factor into our reasoning what we choose to pay attention to.

Propositions
In much of our reasoning we take into consideration some aspect of propositions besides truth-value and form. For example:

- subject matter
- the constructive mathematical content of a proposition
- the likelihood of a proposition being true
- the connotation or sense of a proposition
- the ways we could come to know whether a proposition is true

To take account of such an aspect of propositions in our formal models of logic, we should not change our notion of a proposition. Perhaps this other aspect seems significant for only some restricted class of propositions. After all, who would be concerned with the constructive mathematical content of "Ralph is a dog"? We might wish to specify that restricted class when we first set out a logic, but just as often we discover that class by using the logic. We can do so only if we retain the same meaning for those things we call true or false.

Propositions   A proposition is a written or uttered piece of language that we agree to view as being either true or false but not both.[1]

You may balk here, thinking of some of the aspects listed above. Why should a complex assertion be true or false? Why not simply nonsensical, or meaningless, or unacceptable? There are only two classes of propositions: those that are true, that correspond to the case, part of which is how we understand the connectives; and those that are not true. Third truth-values, undefined truth-values, dual truth-values, levels of plausibility, all these can be taken into account as the content of a proposition. There are only two mutually exclusive truth-values, and all logicians ascribe to something like this view: in the end, they parcel out propositions into those that are acceptable to proceed on as the basis of reasoning in determining what is the case, and those that are not.[2]

We have or adopt a background of agreements. Whether we do so for metaphysical, psychological, physical, pragmatic, or other reasons we might not know. But given such a background, our notion of truth and our analysis of inferences is not a matter of agreement.[3]

If you wish, you can view this dichotomy of propositions as a simplification, a simplicity constraint. But if it is, then it seems to be one that is embedded in the way we (currently) perceive the world.

The logical connectives

Almost all reasoning involves the ordinary language connectives *and, or, not,* and *if . . . then . . . .* We expect to see an account of them in each analysis of how to reason well.

In giving such an analysis we abstract from our use of these ordinary connectives in order to pay attention to only truth-value and perhaps an additional aspect of propositions. We replace the informal connectives with formal ones. Let's use these symbols for the formal replacements:

∧	*conjunction*	¬	*negation*
∨	*disjunction*	→	*conditional*

The interpretations of these will be different according to whether or which aspect of propositions besides truth-value we are concerned with. But then shouldn't we use different symbols for each different interpretation?

By using the same symbols, we will be able to compare how various semantic assumptions affect our logics. And the comparisons are apt because these aren't different notions of "if . . . then . . ." or the other connectives but the same notion taking into account different aspects of propositions. We aren't talking different languages; we're paying attention to different aspects of our speech.[4]

So for each formal logic we should have an explanation of these four connectives, though by defining some in terms of the rest fewer may be taken as primitive. Other connectives may be important for a particular aspect of propositions we are studying, and the general semantics below can be extended to accommodate them.

To give the forms of the propositions we will study, we establish a formal language $L(p_0, p_1, \ldots \neg, \rightarrow, \wedge, \vee)$ in the usual manner, using the propositional variables $p_0, p_1, \ldots$.[5]

No formula of the formal language, such as $p_0 \wedge p_1$, is true or false. It is only when we assign propositions to the variables, such as "Ralph is a dog" to p_0 and "$2 + 2 = 4$" to p_1, and explain how we will understand the formal connectives, do we have a proposition, a formula of the *semi-formal language*, that is true or false. Our intention is to assign to the variables propositions that have no internal structure relative to the connectives we are examining; they will be *atomic*.

It is not for the logician to say which atomic propositions are true nor to say what content or relative aspect each atomic proposition has. Given that each atomic proposition under consideration has a truth-value and some property relative to the aspect of propositions under consideration, our job is to evaluate the truth-value and aspect of compound semi-formal propositions and to show under what circumstances some propositions follows from others.

Compositionality and the Division of Form and Content

In a model the only properties of a proposition that matter are its form, its truth-value, and its content relative to the one aspect besides truth-value that is being considered. Therefore, the truth-value of a compound proposition must be a function of these properties of its parts. If not, what else would determine it? If something transcendent occurs when a connective joins two propositions, how are we to reason? Reasoning requires some regularity. Or at least to simplify our models we will impose some regularity.

Compositionality The truth-value of a compound proposition is determined by its form and the truth-values and contents of its parts.

Here "parts" is to be understood as "proper parts" and not the whole.

The truth-value of a compound proposition will depend on its form. But all the semantic properties of its constituents should be accounted for by their truth-values and contents. Certainly that's so when the constituents are atomic. But to ensure that this is also the case when the constituents are compound, we have to make a further assumption.[6]

The Division of Form and Content If two propositions have the same semantic properties, then they are indistinguishable in any semantic analysis regardless of their form. So if A is part of C, then the truth-value of C depends only on the semantic properties and not the form of A, except insofar as the form of A determines the semantic properties of A.

With these two assumptions, given any compound proposition C in which a proposition A appears, if B has the same semantic properties as A and we replace A by B everywhere in C, then the resulting proposition will have the same truth-value and content as C.

These two assumptions also guarantee that the connectives will operate semantically as truth-and-content functions regardless of whether their constituents are compound or atomic. It only remains to decide which truth-and-content functions correspond to the connectives. Let's start with the simplest case.

Classical propositional logic

We have agreed that all propositions have form and truth-value. If we ignore all other aspects, we are making the *classical abstraction*. This abstraction will yield the simplest symbolic analysis we can devise for our propositional language and notion of a model. In it, the connectives must be evaluated as functions of the truth-values of the parts of the compound.

It is simple to formalize our use of "and" in our reasoning: A∧ B is true just in case both A and B are true. The following table illustrates that choice:

A	B	A∧ B
T	T	T
T	F	F
F	T	F
F	F	F

We use "or" in two different ways in our reasoning. In the *exclusive* sense for "A or B" we mean "A or B but not both A and B." The *inclusive* sense of "A or B" is "A and/or B." It is traditional to choose the inclusive sense to be formalized, as presented in this table:

A	B	A∨ B
T	T	T
T	F	T
F	T	T
F	F	F

In the analysis here, as in all other logics I know, the exclusive sense of "or" can be formalized by using this connective and a combination of the other connectives.

For "not," as in "it's not the case that," the choice for formalizing is easy: ¬A is true just in case A is false, as illustrated in this table.

A	¬A
T	F
F	T

The only non-obvious choice is the formalization of "if . . . then . . .". The table we use is:

A	B	A→ B
T	T	T
T	F	F
F	T	T
F	F	T

Why do we choose this table? If A is true and "if A, then B" is true, we should be able to conclude that B is true. That is the rule of *modus ponens*, which determines the first two rows. But why should A→ B be true in the last two rows? Suppose Dr. E says to Suzy,

If you get 90% on the final exam, you'll pass this course.

It's the end of the term. Suzy gets 58% on the final. Dr. E fails her. Can we say that Dr. E lied? No. So the claim is true, even though the antecedent is false and the consequent is false (the fourth row). But suppose Dr. E relents and passes Suzy anyway. Can we say he lied? No, for he said "if " not "only if ." So the claim is true, even though the antecedent is false and the consequent is true (the third row). The formalization of "if . . . then . . ." in this table is the best we can do when we adopt the classical abstraction. We deal with cases where the antecedent "does not apply" by treating the claim as vacuously true.

To use this analysis, we first assign atomic propositions to the variables of the formal language. Then we assign to each atomic proposition a truth-value; collectively we label the assignments v. The truth-tables then determine the truth-value of every compound proposition formed from the atomic ones. This is what we call a *model*.

This is the simplest formal analysis of how to reason well with propositions as wholes. For any other formal model of reasoning with propositions as wholes, we factor into our analysis (at least) one other aspect of propositions that we consider to be significant.

Set-assignment semantics

To see how to factor into our analysis some additional aspect of propositions, let's begin by viewing that aspect in terms of the content of propositions. For example, the content could be the ways we could come to know whether the proposition is true or the subject matter of the proposition.

We start as before with an assignment of propositions to the propositional variables of our formal language. Then we choose a collection S which has (or represents) the least bits of *content* in terms of the aspect that is being formalized. The collection S should have at least one bit of content or our notion of content will be vacuous. Then every proposition, atomic or compound, is assigned a collection from within S, where we use s to denote the assignment. Often we specify $s(A)$ for every A, and then the collection S is given implicitly as the collection of all the bits in all those particular assignments. Depending on what aspect is being formalized, there may be only one **S** for all models that involve just these atomic propositions or there may be a range of specified collections **S**.

The relation between the content of atomic propositions and the content of compound ones formed from them may be an inductive functional relationship but need not be. When it's not, it's typically because the content in a model is given in terms of some more fundamental notion. For example, ways in which we can come to know that an atomic proposition is true provide a regular basis for content in intuitionistic logics. It is not clear whether it is appropriate to use a notion of content in our reasoning that cannot be reduced to a regular relation of whole to parts.

We must also assign to each atomic proposition a truth-value, the assignment again being denoted by **v**. We need not assume that **s** is independent of **v**. In some logics it is, for example, relatedness logic. In some logics it is not, for example Heyting's intuitionistic logic.[7]

It only remains to decide which truth-and-content functions correspond to the connectives in order to have a model. The conditional is the heart of the matter.

Where there is the correct "connection of meaning" between antecedent and consequent, then only the truth-values of those matter; in that case, the classical table is appropriate for the same reasons that it is appropriate in classical logic. Where there fails to be the connection, then the conditional is false, for we may not infer truths from it.

What is this connection? It will vary depending on the aspect being modeled. For example, in subject matter relatedness logic we require that the antecedent, A, and consequent, B, have some subject matter in common, and hence the relation is $s(A) \cap s(B) \neq \varnothing$. So we could have a model in which "Ralph is a dog $\rightarrow$ 2 + 2 = 4" is false since antecedent and consequent have no common subject matter. For intuitionistic logic the connection is that the ways in which we could come to know the truth of the antecedent are all ways we could come to know the truth of the consequent, so the relation is $s(A) \subseteq s(B)$. For a logic of equality of contents designed as a logic of sense and reference, the relation is $s(A) = s(B)$. In every case the connection is taken to be some fixed relationship, **B**, on contents of propositions. Rather than writing **B**(**s**(A), **s**(B)), we can write **B**(A, B) since it will be clear from context that we're talking about a relationship between contents assigned to A and B. Thus, the truth-table for the conditional is:

A	B	$B(A,B)$	$A \to B$
any	value	fails	F
T	T		T
T	F	holds	F
F	T		T
F	F		T

Often content is not considered significant in evaluating negations, so the classical table for ¬ applies. However, in some logics the content of a proposition must have some particular property, which we can symbolize by **N**, for its negation cannot to be true. For example, in Heyting's intuitionistic logic the property is $s(A) = \varnothing$. So, writing **N**(A) for **N**(s(A)), the general table for negation is:

A	$N(A)$	¬A
any value	fails	F
T	holds	F
F		T

Generally conjunction is interpreted classically. But we may wish to take the contents of the conjuncts into consideration. For example, we can take the content of an atomic proposition to be the time of which it is meant to be true. Then taking the connection between propositions to be that the time assigned to the first comes before the time assigned to the second, the formalizations of "Dick took off his clothes and went to bed" could be true while the formalization of "Dick went to bed and took off his clothes" would be false. So we have the general table for conjunction, where $C(A, B)$ is short for $C(s(A), s(B))$, a relation between contents of propositions:

A	B	$C(A,B)$	$A \wedge B$
any	value	fails	F
T	T		T
T	F	holds	F
F	T		F
F	F		F

Typically the connective ∨ is defined from the other connectives. When it isn't, its truth-table is governed by a relation **A** between contents of propositions: when the relation holds, the table for classical inclusive ∨ is used; when the relation fails, the disjunction is false.

A	B	A(A,B)	A∨B
any	value	fails	F
T	T		T
T	F	holds	T
F	T		T
F	F		F

The relations on contents in the evaluations of the connectives are collectively referred to as the *relations governing the truth-tables*.

Given any assignment of truth-values to the atomic propositions and assignment of contents to all propositions, these tables determine the truth-value of each proposition of the semi-formal language by Compositionality and the Division of Form and Content.

Schematically, then, a *model* for the formal language is:

I

$$L(\neg, \rightarrow, \wedge, \vee, p_0, p_1, \ldots)$$

$\downarrow$

$\{\text{real}(p_0), \text{real}(p_1), \ldots, \text{compound propositions formed}$
$\text{from these using } \neg, \rightarrow, \wedge, \vee \}$

$\mathsf{v} \downarrow$ $\downarrow \mathsf{s}$

$\{T, F\}$ subcollections of **S**
the relations **B, N, C, A**

truth-tables

$\downarrow$

$\{T, F\}$

We've allowed for each of the four basic connectives to depend on only the truth-values of the constituent propositions by allowing the appropriate relation to be universal. However, we have not made provision for models in which the evaluation of $\neg, \rightarrow, \wedge$, or $\vee$ depends solely on the contents and not the truth-values of the parts. The role of such *wholly intensional* connectives in reasoning is not clear. Perhaps for subjunctive conditionals, such as "If dogs could meow, then more cats would be killed," in which a false antecedent does not affect the truth-value of the whole, an intensional interpretation of $\rightarrow$ could be used. A minimal classical modal system, **K**, requires an intensional interpretation of $\rightarrow$ in the form: $A \rightarrow B$ is true iff $\mathsf{s}(A) \subseteq \mathsf{s}(B)$. But there is no indication what reasoning that system might be modeling since the rule of *modus ponens* fails there. A minimal logic of intuitionism also can be given semantics using a wholly intensional connective: $\neg A$ is true iff $\mathbf{s}(A) \subseteq \mathbf{s}(\neg A)$. But there the set-assignment semantics are parasitic on other semantics that are not in keeping with the motivation of the logic. At least one paraconsistent logic has a notion of negation that can be interpreted intensionally, though only in the presence of another negation that is modeled classically.[8] The role of wholly intensional versions of $\neg$, $\rightarrow$, $\wedge$, and $\vee$ as formalizations of our usual understandings of "not", "if . . . then . . .", "and," and "or" needs further analysis.

Relation-based semantics

In formalizing how to reason taking account of subject matter, for example, rather than considering subject matter to be the content of a proposition, made up of some bits, some "subjects," it is more straightforward and perhaps more intuitive to take the relation of one atomic proposition having subject matter in common with another as a primitive relation. We needn't say what the subject matter of "Ralph is a dog" or "$2 + 2 = 4$" is in order to say that those propositions are not related. In that case, we can devise models of our logic by directly stipulating the relations $\mathbf{B}, \mathbf{N}, \mathbf{C}, \mathbf{A}$ on atomic propositions and saying how those can be extended to compound propositions. In a model of type $\mathbf{I}$, we have simply the relations and no set-assignment. Other logics can be presented in this way, too, and when we do so we say that we are giving *relation-based semantics* for the logic.

Abstract models

In any model the only semantic properties of an atomic proposition that we consider are its truth-value and the content assigned to it. So we can abstract our models to:

$\mathbf{II}$ $L(\neg, \rightarrow, \wedge, \vee, p_0, p_1, \ldots)$

$\quad\quad\quad\quad\quad\Big|\quad$ $\mathbf{v}, \mathbf{s}$, and truth-tables governed by $\mathbf{B}, \mathbf{N}, \mathbf{C}, \mathbf{A}$

$\quad\quad\quad\{\mathsf{T}, \mathsf{F}\}$

A formula, understood as representing a proposition, is true or false in this model according to whether $\mathbf{v}(A) = \mathsf{T}$ or $\mathbf{v}(A) = \mathsf{F}$. We have explicitly recognized that all other aspects of the propositions are to be ignored.

Up to this point, models have been built from propositions that are sentences, and content sets are motivated and argued for with respect to the aspect of propositions under consideration, as are the relations **B**, **N**, **C**, and **A**. Even in diagram **II** the assumption is that the model arises from one as in diagram **I**.

We are not obliged to carry out our logical investigations with any greater abstraction than this. But to achieve full generality, to simplify our proofs, and to isolate the structural nature of these models, it is useful and now customary to make a number of assumptions relative to models of type **II**. These assumptions allow us to use mathematics in our studies.

The first assumption we make is to view the set of propositional variables as a completed whole, PV. We also view the collection of well-formed-formulas as a completed totality, Wffs. We treat these collections mathematically as sets. Then we make the following abstraction.

The Fully General Abstraction

1. Any function $\upsilon: \mathrm{PV} \rightarrow \{\mathsf{T}, \mathsf{F}\}$ is suitable to use as a truth-value assignment to the variables.

2. Any set **S** and any function **s** : Wffs $\rightarrow$ subsets of **S** which (together with υ) satisfy the structural rules for modeling the aspect under consideration are suitable to use as a content assignment.

3. Given υ and a pair **s** and **S** as in (1) and (2), any relations **B**, **N**, **C**, and **A** on the sets assigned to the variables that satisfy the structural rules for modeling the relations governing the truth-tables are suitable to use in a model.

To use parts (2) and (3) we need to have proceeded beyond ordinary language examples or the ability to produce models of type **I** to a structural analysis of contents of propositions.

The fully general abstraction obliterates the differences between aspects of propositions that satisfy the same formal rules. For example, the same structural analysis, $\mathbf{s}(A) = \bigcup \{\mathbf{s}(p) : p \text{ appears in } A\}$, is used for the referential content of a proposition and the content of a proposition taken to be its consequences in classical logic.

We could assume the fully general abstraction and use mathematical tools on our models without assuming anything about infinite completed totalities such as PV, Wffs, or **S**. In that case we're making a *Finitistic Fully General Abstraction*.

If we make these abstractions, it's because we believe that the results we obtain using them when applied to propositions in fully motivated models of type **I** won't contradict our original assumptions and intuitions, that is, our semantics. If they do, then we can trace back along the path of abstraction to see how anomalies arise.[9]

Semantics and logics

The Fully General Abstraction reduces the semantics to a translation from one language, $L(\neg, \rightarrow, \wedge, \vee, p_0, p_1, \ldots)$, to another, the informal language of mathematics. That by itself is no semantics at all; it is just a translation. The fully general abstract semantics are rules governing forms of meanings, comparable to an axiomatization of the logic in the more formal language $L(\neg, \rightarrow, \wedge, \vee, p_0, p_1, \ldots)$. Another syntactic characterization from a different point of view does not give meaning.

Semantics are only given when arguments are made, agreements based on common understandings are reached, and our intuitions sharpened so that we can give an analysis in terms of models of type **I**. Questions of meaning, truth, reference, and our relation to "the world" must be dealt with. Otherwise we have no semantics but only an empty formalism. What I have presented here is the framework for semantics for a propositional logic. It is fleshed out into a logic when:

1. We choose a semantic aspect of propositions to formalize or choose to formalize none except truth-value.

2. We explain that aspect in simple terms, giving examples, developing an intuition, justifying the assumption that *a proposition can be viewed as having this aspect*. In doing so we may come to restrict ourselves to a certain class of propositions to which this assumption applies, either ordinary language propositions or ones from some technical scientific language. We may also choose to add further connectives to the formal language.

3. We argue that in the context of the reasoning we are formalizing it is acceptable to ignore all other aspects of propositions, justifying the abstraction that *the only*

*properties of a proposition that matter to (this) logic are
its form, its truth-value, and this one additional aspect.*

4. We stipulate a class of content-assignments that are
appropriate to model this aspect, usually through rules they
must obey; that is, a structural analysis of content is given.
In particular, we decide whether for atomic propositions the
assignment of contents is independent of the assignment of
truth-values.

5. We explain the connections of meaning that govern the
truth-tables and stipulate a class of relations **B**, **C**, **A**, and
property **N** that are appropriate to model them. These classes,
too, are usually presented on the basis of a structural analysis.

6. We choose whether to make the Fully General Abstraction
of models, the Finitistic Fully General Abstraction of models,
or no abstraction at all.

7.   We return to our examples and show that the formalization of them and the resulting semantic analysis are reasonable
in terms of the original motivation (2). When our pre-formal
intuitions clash with the formal analysis we may explain why
our semantics are appropriate to act prescriptively.

The *logic*, then, is this analysis: the syntax and the formal
semantics developed through (1)–(7).

Semantic and syntactic consequence relations

There are two major concerns of formal logic that result in other uses
of the term *a logic*.

As logicians we cannot specify any particular model as being
an accurate interpretation of "the world." We can only say which
compound propositions are true given specific truth-assignments
and content-assignments. However, certain propositions will be true
regardless of the assignment.[10] They are true (in this logic) due to
their (propositional) form only. Their schematic form as wffs will be
assigned a true proposition in every model. These are the "universal
truths" (of this logic). We call formal wffs true in every model
tautologies. The tautologies are the forms of what we must accept
due to our assumptions about how we (should) speak and reason.
When the classification of propositions true due to their form only

becomes the main focus of an investigation, the class of tautologies
is sometimes referred to as the logic.

It is not just what is true that is of interest, but what follows from
what. That is, given certain propositions as premises, what can we
conclude due only to our assumptions about how we (should) speak and
reason? We are concerned with whether a proposition B follows from
the assumption of the truth of a proposition A or of a collection of
propositions Γ. Given a logic L, we write $A \vDash_L B$ to mean that in every
model of L in which A is true, so is B. We write $\Gamma \vDash_L B$ to mean that
in every model in which every A in Γ is true, so is B. This is the
semantic consequence relation for L. In this notation $\vDash_L A$ means
that A follows from no propositions, and so is true in all models; that is,
A is a tautology. The relation $\vDash_L$ is a formalization, for this logic, of
one proposition *following from* one or more other propositions.

Given a class of models and associated notion of tautology, there
may be other ways to define a semantic consequence relation peculiar
to the logic in question. A different notion of semantic consequence for
subject matter relatedness logic results in a paraconsistent logic[11]; T. J.
Smiley discusses three different consequence relations for a particular
many-valued logic.[12] Hence, the term *a logic* is sometimes reserved
for the chosen semantic consequence relation.

There is a another conception of logic concerned with form and not
meaning. A logic is characterized as a formal system of axioms and
rules meant to formalize the notion of proof. Then one proposition A is
taken to be a consequence of a collection of wffs Γ if it can be derived
by the rules from Γ and the axioms.[13] However, to devise a logic
entirely syntactically is to reject any attempt to make explicit the
semantic assumptions we use in justifying our reasoning.

The use of these general semantics

The general form of semantics I've presented here is based on a view of
propositional logics as a spectrum, all of the same general form, each
based on some aspect of propositions in addition to form and truth-
value, with the exception of classical logic which ignores all other
aspects. As we vary the aspect, we vary the logic. What we pay
attention to in our reasoning matters.

To establish that these semantics do indeed yield a structural and
conceptual overview of many logics is one of the goals of *Propositional Logics*. There I show that analytic implication, intuitionistic

logics, many-valued logics, and other logics can be presented in terms of this overview, giving semantics within the general framework that conform to and reflect the intuitions of the originators of each logic. Such semantics are not meant to replace the semantics devised by the originators of the logics. For example, under certain assumptions, possible-world semantics are a good model of the ideas of intuitionistic logics. But providing uniform semantics that are in reasonable conformity with the ideas on which the various logics are based allows for comparisons and gives a uniform way in which to approach the sometimes overwhelming multiplicity of logics. In particular, it gives us a way to understand, categorize, and answer questions about translations between logics, as I show in "On Translations" in this volume.

It is equally important that this general form of semantics provides a simple tool for incorporating into logic many different aspects of propositions that for the most part have been treated only informally. We now have a framework in which to formalize, discuss, and compare different notions as they affect our analyses of reasoning well. For example, we now have a semantic analysis of "relevance" in terms of subject matter. And in *Propositional Logics* I use this general form of semantics to develop a logic that incorporates a notion of referential content of a proposition. In "A Propositional Logic of Temporal Connectives," Esperanza Buitrago-Díaz and I use this method to show how to formalize reasoning with connectives like "before" and "after," taking the content of a proposition to be the time(s) of which it is meant to be true.

The referential content of a proposition, however, is not a primitive notion. It depends on the referential content of the predicates and names of which the proposition is composed. The internal structure of propositions matters, and predicate logic is a further test of the aptness of these ideas.[14]

The semantic framework that I have set out can be taken as a very weak general form of logic, a general form that becomes usable only upon the choice of which aspect of propositions we deem to be significant. But then is logic relative to the logician? That's what the next essay, "Why Are There So Many Logics?", is about.

Appendix   Quine on the unity and division of logics

The views of W. V. O. Quine on logical connectives set out in Chapter 6 of his
Methods of Logic are in marked contrast to those in this essay.

Suppose we were to say that subject matters affect conjunction. And
suppose further that we deny that "Ralph is a dog ∧ 2 + 2 = 4" is true, while
assenting to each conjunct. Quine, I believe, would argue that the use of the
symbol "∧ " is inappropriate.

> If a native is prepared to assent to some compound but not to a
> constituent, this is a reason not to construe the construction as
> conjunction. p. 82

I assume Quine would say similarly that if "the native" were to assent to both
constituents but not the compound, we should not construe the construction as
conjunction.

Certainly we would no longer be talking of classical conjunction. But
Quine stresses that we are, first of all, discussing an ordinary language. No
native speaks using ∧ , ⌐ , → , or words like "conjoin," unless he or she is a
logician. People speak using words such as "and" or "et." Moreover, no one
invariably uses "and" truth-functionally. Yet it would only be on the basis of
everyone doing so that Quine could claim that logical truths are obvious:

> Naturally, the native's unreadiness to assent to a certain sentence
> gives us reason not to construe the sentence as saying something
> which would be obvious to the native at the time. . . . I must stress
> that I am using the word "obvious" in an ordinary behavioral sense,
> with no epistemological overtones. When I call "1 + 1 = 2" obvious
> to a community I mean only that everyone, nearly enough, will
> unhesitatingly assent to it, for whatever reason; . . . Logic is
> peculiar; every logical truth is obvious, actually or potentially. p. 82

Simple logical "truths" involving the formal "∧ " that we explicitly agree
to interpret truth-functionally are certainly obvious, but not in the sense of
Quine, for no one speaks a language with that "word" in it. Logical truths,
whatever they may be, are not obvious in his sense if it means only that, nearly
enough, people will unhesitatingly assent to them. That is because "nearly
enough" conceals an immense problem of explication; for example, we know
that "or" is not used nearly always in the truth-functional inclusive sense.
Moreover, subjects of a tyrant may unhesitatingly assent to "The tyrant is
good." Logic is not so peculiar; the reason for assent matters.

There is evidence in our language and reasoning to support many different
models of the same connectives. We choose the connectives to investigate and
build our formal language first, then decide whether to use the classical interpretation. We are speaking the same language; we are paying attention to

different aspects or uses of it. There is a change of subject in that sense, but we may still have real disagreements about which model represents the way people do reason (if you're listening to the natives) or should reason. Quine's sense of obviousness is accommodated and better understood as agreement between, archetypally, two participants in a logical discussion.

Notes

1. (p. 32) See "Truth and Reasoning" in this volume for a discussion of why this definition can serve those who work with any of the propositional logics discussed in this essay. In other contexts I call these "claims."

2. (p. 32) See "Truth and Reasoning" in this volume for a fuller discussion.

3. (p. 32) Compare C. I. Lewis and C. H. Langford in *Symbolic Logic*:

> We make an inference upon observation of a certain relation between facts. Whether the facts have that relation or not we do not determine. But whether we shall *be observant of* just this particular relation of facts, and whether we shall *make that relation the basis of our inference* are things which we do determine. p. 258 (italics in original)

4. (p. 33) Even the crows on the roof tops are cawing about the question which conditionals are true.

> Callimachus , from *Adv. Math*. 309–310 quoted in Mates, *Stoic Logic*, pp. 42–43

5. (p. 33) See "Valid Inferences and Possibilities", p. 6, for a definition.

6. (p. 34) See "The Error in Frege's Proof that Names Denote" in this volume for an example of what can go wrong if we don't adopt this assumption.

7. (p. 37) See *Propositional Logics* for these and the other examples discussed in this paper.

8. (p. 40) See "A Paraconsistent Many-Valued Logic J_3" by Itala D'Ottaviano and me, revised as Chapter VIII of my *Propositional Logics*.

9. (p. 42) See "Models and Theories" in *Reasoning in Science and Mathematics* for a discussion of modeling and abstraction in science and logic. The role of mathematics in analyses of reasoning is discussed in *Propositional Logics* and *Predicate Logic*.

10. (p. 43) This does not hold for some many-valued logics; see *Propositional Logics*.

11. (p. 44) See my "Paraconsistent Logics with Simple Semantics" or Chapter VIII of *Propositional Logics*.

12. (p. 44) "Comment on 'Does many-valued logic have any use' by D. Scott."

13. (p. 45) See Chapter II.L.1 and II.N.1 of *Propositional Logics* for a brief history. In Chapter II.K.3 of *Propositional Logics* these ways to present a logic are summarized.

14. (p. 45) I set out the general form of this semantic overview for predicate logics in *Predicate Logic*. In "Relatedness Predicate Logic" Stanislaw Krajewski and I give a predicate logic of subject matter relatedness logic.

Why Are There
So Many Logics?

If logic is the right way to reason, why are there so many logics?

The choice of a logic depends on what we pay attention to in our reasoning. Calling one logic right and another wrong often arises from judging one by the background assumptions of the other. Necessity in our reasoning, if there is any, is in the common background for all logics.

It seems that we can understand all logics in the same way: we start with our everyday language and abstract away certain aspects of linguistic units and take into account certain others by making idealizations of them. The aspects we pay attention to determine our notion of truth.[1]

There does not seem to be a difference between logical and pragmatic aspects of what we call propositions, not at least any difference we seem to be able to justify. What direct access have we to the world but our uncertain perceptions? And how can two of us share exactly the same perceptions or thoughts? Vagueness seems essential to communication. So to call a sentence true seems at best a hypothesis we hope to share with others. This sharing, which in a sense amounts to objectivity, is brought about by common understandings, which I call agreements. But this notion of truth is so basic to our experience and the fit of thought to the world that we can no longer allow ourselves to see that truth is in how we abstract, perceive, and agree, lest we have no language to talk.

The word "agreement" is wrong, and "convention" even more so. As John Searle points out, we don't have good nonintentional words with which to describe our backgrounds.[2] Almost all our conventions, agreements, assumptions are implicit, tacit. They needn't be either conscious or voluntary. Many of them may be due to physiological, psychological, or, perhaps, metaphysical reasons; for the most part we do not know. Agreements are manifested in lack of disagreement and the fact that people communicate. To be able to see that we have made (or been forced into, or simply have) a tacit agreement is to be challenged on it. If the assumption is sufficiently fundamental and widely held, we call the challenger "mad."[3]

The explicit background is quite different from the implicit, though I use the same word "agreement" for each. I think that's the best term, for it allows us to use the word "disagreement" when our backgrounds clash. It points to the background as it affects our interactions with one another. And it is the drawing of the implicit background into explicitness by abstracting, idealizing, and simplifying that is the basis of formal logic.

It may be wrong to ascribe a uniformity to our backgrounds. As Peter Eggenberger has pointed out to me, from the comparatively uncontroversial "For every act of communication there are some agreements on which it is founded," it does not follow that there are some agreements on which every act of communication is founded. But I have tried to show that this latter assumption is reasonable at least with respect to logical discourse.[4]

In a conversation, John Searle tried to convince me that truth and referring can't be a matter of agreement. He argued that if he says "The moon is risen" and I say "Which moon?"' then he has successfully referred even if I keep saying "Which moon?"

But I believe it is a matter of agreement that we say the same object is in the sky each night, not 28 different ones, or one new one for each night of eternity, or that it is an object and not a process. There may indeed be only one object "out there," but how can either of us ever know that with any certainty if that transcends our perceptual framework, our background? Even relative to our background, anything that is beyond direct immediate experience must be a matter of agreement for referring, while much that is of direct immediate experience is so theory-laden as to be called a matter of agreement for referring, too.

But, Peter Eggenberger argues, suppose we're playing a game of chess. The rules are completely explicit: a rook moves in this fashion, pawns in that. The game ends with either a checkmate, a stalemate, or an agreed draw, and so on. Now suppose you move and announce, "Checkmate." Or as he puts it, suppose I'm in a position that "says" I'm in checkmate. Then I can't get out of it by saying that I disagree.

But of course I can, though it's extremely unlikely (it seems to us) that I could do so in good faith, that is, really not recognize it as checkmate by your "objective" standards. If I do say "I don't agree," either through perversity or my actually not regarding such a position as checkmate, then pretty soon I'll find no one to play chess with me. Perhaps that's not such a great loss. But if I do not agree with the community's language agreements and assumptions, then I cannot get anyone to talk with me. That is a loss. Objectivity arises because we'd all go mad (or be mad) if we didn't act in conformity with some implicit agreements and rules. We are all built roughly the same, and we have to count on that in exchanging information about our experience. Given a particular shared background, there will be plenty of room for experience and facts.

When we adopt a language, we can't help but adopt the agreements on which it is based. For example, consider what happens when I teach real analysis, the theoretical foundations of calculus. I do not believe that there are any infinite entities, or at least none corresponding to what we call "the real numbers." But when I teach real analysis, I have to adopt the language of the classical mathematician. I could preface each remark of mine with "Of course, this is assuming that infinite totalities exist, which I don't really believe." But I don't; I make that comment at the beginning of the course and then, slowly, forget it. I have to in order to be able to talk in the language of real analysis, a language that has grown out of human experience and is therefore accessible to me. I am only able to talk coherently in that language if I accept its background assumptions. The more I talk the language, the more likely I am to forget that my acceptance was hypothetical. I convince myself that I have not betrayed my beliefs by saying that the theory I am teaching is an idealization of experience. But it could be said that I have not betrayed my beliefs in the same way when I learned Polish and began to speak it in Poland and in doing so adopted its assumptions and conventions. I have not forgotten the categories of the world of my language; I have just put them aside in order to communicate with people who do not have or use them.

But I mustn't be fooled into thinking that a logic or language can be justified by its utility. An example comes from a paper by Michael Wrigley and a discussion I had with him.[5]

We use Peano Arithmetic, **PA**, because we believe it's "right." Now suppose we encounter someone who used a deviant arithmetic, **DA**, which we see is inconsistent. Alan Turing said that we could distinguish the two arithmetics because if we built bridges using **DA** they'd fall down (more often). Well, in a sense he's right and in a sense he's wrong.

We ought to suppose that the practitioners of **DA** have their own background assumptions, which surely must differ from ours, for with our background if we used **DA** to build bridges, they would fall down more often. But why shouldn't we assume that relative to their background assumptions the practitioners of **DA** would build perfectly fine bridges? After all, **DA** would reflect their background assumptions well or they wouldn't use it. There is a temptation here to say that we do not have to account for the deviant arithmetic because it's no arithmetic at all: they must not be talking about the natural numbers. But we were assuming that we had some good reason for calling it "arithmetic" in the first place. Disagreements do not disappear by saying that the subject has been changed.[6] Moreover, how do we know that even among ourselves we all understand **PA** in the same way, that there is a common subject? We are back again at the not unreasonable hypothesis that we have a shared background so that we can communicate. And only relative to a particular background do criteria of utility have force. Turing was trying to fill in the dotted line:

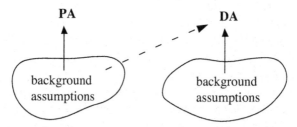

As Benson Mates pointed out to me, often enough it seems you don't believe it because it's useful; it's useful because you believe it. There's pragmatic value in believing that it's not a matter of pragmatics. Even to think that our fundamental background assumptions could be challenged is paralyzing to most users of language.

Well, then, why should we question them? Because we have disagreements. There are many logics. It isn't a "let's pretend" situation, as with the example of deviant arithmetic. The objectivity of our backgrounds really has been questioned.

Yet the classical logician, the intuitionist, the modal logician, the practitioner of a relevance or many-valued logic all do communicate and work together. They write journal articles and read ones written by the others, they collaborate, however uneasily, they discuss. One possible explanation of this is that the usual daily languages of all these people share so many background assumptions that the practitioners of, say, intuitionism, cannot free themselves of the classical way of thinking any more than I can free myself of the classical real analysis way of thinking when I am helping my mathematics students. Perhaps intuitionism cannot be the challenge the intuitionists intended it to be because they only adopted a new way of mathematically talking (and thinking) while retaining the language and assumptions of the culture in which they live.

The relevance of this explanation to doing logic and seeing the unity of logics is that because the practitioners of the various logics share the same background assumptions and agreements that come with the use of Western languages and culture (or perhaps that come with any human culture), those assumptions must be evident in their work. Those more fundamental assumptions are what I explicitly use in setting up the structural overview of propositional logic I've presented: there is a common notion of a smallest linguistic unit, called "a proposition," that can be called true, and common notions of the connectives as portrayed in the general truth-tables. Those tables reflect that some notion of meaning or content is ascribed to propositions, and a connection of meanings or contents must be made for a compound proposition to be accepted. If the connection is there, then only the truth-values of the constituents remain to be considered, and we evaluate the connectives according to the standard classical tables. Always present is the Yes–No dichotomy that we impose on (or is imposed on us by) experience.

The differences between the logics must be superficial relative to these assumptions, and the differences are, as I see it, in the choice of which, if any, less fundamental aspect (or possibly aspects) of propositions other than truth-values are to be taken into account in reasoning. Even then, some aspects are so close to fundamental, so near the heart of the background, that a challenge to them seems incomprehensible to most of us. Such aspects or notions are the ones we tend to label "logical." To me, whether a notion is called logical is a measure of how fundamental it is to our reasoning, our communication, our view of reality, not how close it is to reality. The following diagram pictures this view.

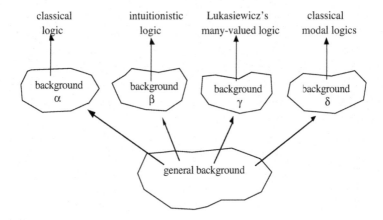

For someone with background α , classical logic seems objective:
we must reason in accord with it. And that is because with that background it seems inconceivable to act otherwise. But it is not inconceivable to have another background, as I've argued, only extremely
difficult to enter into one.

If there is a nonrelative sense of "must" in this picture, it would
seem to be in the general background. Nothing I have done depends
on whether the general assumptions on which these logics are based
are necessary truths. The metaphysical arguments may begin again,
only relocated. We don't get rid of the background, for language must
be anchored to the world. We only show that more or less of it as
represented in language and logic can be considered universal.

Yet all our formalizations, all I've done, are false if taken to be
exact representations of our implicit backgrounds. What is thought
remains thought, what is the external world remains the external world,
and what we say is and can only be what we say, either about, because
of, or through the world and thought. It seems to me false that what is
said can be directly of these other realms or a perfect representation of
them. The explicit agreements we codify are abstractions, idealizations, simplifications, and hence a distortion of our real background.
These explicit agreements may make it easier for us to communicate,
and we may henceforth act in accord with them; and doing so can
sometimes feed back into our backgrounds becoming imperceptibly,
in a new nonexplicit way, our unconscious, implicit background.

It seems hard for me to conceive of someone who reasons who
doesn't have a smallest unit of language to which a Yes–No dichotomy

is applied, and hence to conceive of someone who uses connectives that would not be expressible by the tables of the general semantics for propositional logics. So perhaps the story of propositional logics I give is universal. For it to be useful for us, for us to develop it, we will probably have to believe that.

Notes

1. (p. 49) A presentation of the formal analysis of logics according to this viewpoint can be found in the accompanying essay "A General Framework for Semantics for Propositional Logics" in this volume.

2. (p. 49) *Intentionality*, Section V of Chapter 5.

3. (p. 50) Compare Benjamin Lee Whorf, "Science and Linguistics",

> We cut nature up, organize it into concepts, and ascribe significances as we do, largely because we are parties to an agreement to organize it in this way—an agreement that holds throughout our speech community and is codified in the patterns of our language. The agreement is, of course, an implicit and unstated one, BUT ITS TERMS ARE ABSOLUTELY OBLIGATORY; we cannot talk at all except by subscribing to the organization and classification of data which the agreement decrees.
> p. 213–214 (capitalization in original)

4. (p. 50) See again "A General Framework for Semantics for Propositional Logics" in this volume.

5. (p. 51) "Wittgenstein on Inconsistency."

6. (p. 52) Compare the discussion in the appendix to "A General Framework for Semantics for Propositional Logics" in this volume.

Truth and Reasoning

A major goal of reasoning is to establish truths and to determine what would follow if certain assumptions are true. There are many different notions of what is true, both in what kinds of things are true or false and what makes them true or false. By looking at what is common to those notions, we can find an idea of truth and the things that are true that can accommodate many particular views of truth and account for the wide agreement on what counts as good reasoning.

What kind of things are true or false?

We reason to try to determine what is true and what is false. We reason to try to determine what would be true if certain assumptions are true. We learn to reason as we learn our language. We reason together so we can reason alone. Reasoning is a kind of communication.

To reason well we need to understand the nature of truth and falsity and how those classifications function in our reasoning. To begin, let's consider what kind of things are true or false. When we reason together, we use language. We make sounds or write inscriptions. Or, reasoning with ourselves, we imagine a sound or inscription.[1]

Whether we view such utterances and inscriptions as true or false, or whether we think it's the meaning of such language that's true or false, or whether we think such language points to an abstract realm of things that are true or false, it's through our use of language that we reason. So let's begin by focussing on such utterances and inscriptions. Later we'll consider whether they only represent or point to what is actually true or false.

What kinds of linguistic sounds and inscriptions are (or represent what is) true or false? It could be just a word, as when looking out the window I say to a friend, "Raining." It could be as complex as a sentence that takes half a page. It could be an equation composed of mathematical symbols.

We can, however, exclude questions, commands, and wishes. We can also exclude sounds that are nonsense, such as "Frabjous day," or apparently good language that is really meaningless, such as "7 is divisible by light bulbs." We can exclude ambiguous utterances, such as "I am half-seated," and ones that are too vague, such as "America is a free country."

It would seem, then, that for a piece of language to be a candidate for being true or false, it must be completely precise and intelligible. But if only language that is completely precise can be true or false, then "Strawberries are red" does not qualify. Which strawberries? What hue of red? Measured by what instrument or person? So we couldn't analyze:

(‡) If strawberries are red, then some colorblind people cannot see
        strawberries among their leaves.
     Strawberries are red.
     Therefore: Some colorblind people cannot see strawberries
        among their leaves.

This is an example of acceptable reasoning, reasoning that any rules we devise about how to reason well must be able to account for.[2] Yet any attempt to make the language in (‡) fully precise will fail. At best we can redefine terms, using others that may be less vague.

No two people have identical perceptions, and since the way we understand words depends on our experience, we all understand words a little differently.[3] There has to be some wiggle room in the meaning of what we say for us to be able to communicate. When Zoe says, "Spot's barking woke Dick" (she was up late working on her term paper when Spot started yelping), is that true or false? The language is not too vague, unless we need, for some reason, some purpose, more precision. When did Spot begin to bark? When exactly did Dick awake? How long did Spot bark? How loud? What kind of yelping or howling or growling or ferocious arfing was it? "Oh, that doesn't matter," says Zoe, "you know very well what I mean." The issue isn't whether a sentence or piece of language is vague but whether it is *too vague*, given the context, for us to be justified in saying it's true or false.

Our goal is to have complete clarity in our reasoning. But that is and can only be a goal. What we take to be true or false must depend to some extent on our purposes, on what we pay attention to. Every description of the world of our experience is at best partial. We are limited in our descriptions by the resources of our language, by the resources of our sensory apparatus (how do dogs smell so much?), by how we process the experiences of our senses, and by how much of what we process we pay attention to. Do we choose to pay attention to only part of our experience? Psychologists will tell us no, yet we have available to us more than we first note. A clock striking four o'clock may not register as four to us; but if asked to recall, often we can say, "One, two, three, four. Yes, it struck four times."

Meaning does not reside in a piece of language; it does not reside in us; it does not reside in the world. It resides in us, language, and the world—in us using language to talk about the world and our experience. Meaning is in a particular use of language, in a particular context, among particular people. Meaning is not fixed, not for us individually and most certainly not among ourselves when we talk. We have to negotiate meaning.

We negotiate meaning to try to understand each other better, or perhaps at all. I, you negotiate meaning with ourselves each time we use language in a different way, or in a different context, or just when we reflect on what we say. We negotiate meaning with others, trying to fix more closely how we understand what we say so that we can have some confidence that we are communicating, that we understand together. The need for such negotiation may be evident only from our actions and disagreements. When we negotiate meaning with ourselves, we may do no more than think about what we are saying.

Whether we accept a piece of language as a candidate for being (what represents what is) true or false depends on our purposes, which we may negotiate as we negotiate meaning. What we pay attention to may be culture-bound by the resources of our language and by what we deem important in our experience, which is often codified by our language. But once given that framework in which to deal with our experiences, truth need not be relative.

"No, Spot's barking didn't wake Dick since he was stirring and coughing before Spot started to bark." We can claim evidence. We can compare experiences within our framework. But no framework is available to us to judge all of what there is, for we are limited. And thank Dog for that, for what horror to perceive all undiluted, to register all we perceive, to put into language all we have registered.

Perhaps there is a real framework in which to judge truth and falsity, in which to judge all. Perhaps God, or the gods, or Dog who smells all can perceive such a framework. Perhaps we can or should hold up such a framework as what we strive to see. But we are investigating how to reason well, and our limitations cannot be ignored.

So it is sufficient for our purposes in reasoning to ask whether we can agree that a particular piece of language, or class of inscriptions as in a formal language, is suitable to assume to be true or false, that is, to have a truth-value. If we cannot agree that a particular sentence such as "The King of France is bald" has a truth-value, then we cannot reason together using it. That does not mean that we employ different methods of reasoning or that all reasoning is psychological; it only means that we differ on certain cases. The assumption that we agree that a piece of language has a truth-value, that the imprecision of it is inessential, is always present, even if not explicit.[4]

The words "agree" and "negotiate" are somewhat misleading. Almost all our agreements, conventions, assumptions are implicit, tacit. They needn't be conscious or voluntary. Many of them may be due to physiological, psychological, or perhaps metaphysical reasons; for the most part we don't and perhaps may never know. Agreements are manifested in lack of disagreement and in that we communicate. Agreements are the result of our negotiations, which are often done tacitly with no verbal assent or even talk. To be able to see that we have made, or been forced into, or simply have an agreement is to be challenged on it. If I say "Cats are nasty" and you disagree with me, then I know that you consider that utterance to be true or false.[5]

So I suggest that we begin with the following definition:

A *claim* is a written or uttered piece of language that
we agree to view as being either true or false.

When reasoning with ourselves, we might not write or utter anything. But such reasoning can be understood as dependent on how we reason with sounds and inscriptions, using our imagination to think of the claims.[6]

Perhaps it is these utterances and inscriptions that are true or false
in the negotiations and agreements we have made. Perhaps it is the
meanings of them that are true or false in the negotiations and agreements we have made. Perhaps these utterances and inscriptions only
represent or point to some abstract things called "propositions" that are
really true or false, the pointing being what we negotiate and come to
agreements about. No matter. It is these we use when we reason
together. So it is these we can discuss, leaving to the metaphysicians
to clarify the grounds of our negotiations and agreements.

Example I wish I could get a job.

Analysis If Maria, who's been trying to get a job for three weeks,
says this to herself late at night, then this isn't a claim. It's more like a
prayer or an extended sigh.

But if Dick's parents are berating him for not getting a job, he
might say, "It's not that I'm not trying. I wish I could get a job."
That might be true, but it also might be false, so the example would
be a claim.

It is not a sentence or an inscription devoid of context that is a
claim. A claim is a specific piece of language in a specific context.

Example Wanda is fat.

Analysis Wanda weighs 120 kgs and is 1.7m tall, so almost all of
us would agree with this example. We take it to be a claim.

If Wanda were the same height yet weighed only 50 kgs, we'd all
disagree with it, which shows that we'd take it to be a claim then, too.

But if she weighed 72 kgs, we'd be unsure. It isn't that we don't
know what "fat" means; we just don't think the sentence is true or false.
We don't classify it as a claim. It isn't that our notion of claim is
vague; we're clear that we won't classify it as true or false. It's the
sentence itself that is too vague in that context to classify as a claim.

But if there's no clear division between what we mean by someone
being fat or not being fat, doesn't that mean that we don't have a clear
notion of claim? No. That we cannot draw a line does not mean there
is no obvious difference in the extremes.

Claims are types

Suppose we're having a discussion. An implicit assumption that
underlies our talk is that we will continue to use words in the same way,
or, if you prefer, that the meanings and references of the words we use

won't vary. This assumption is so embedded in our use of language that it is hard to think of a word except as a representative of inscriptions that look the same and utterances that sound the same. I do not know how to make precise what I mean by "look the same" or "sound the same," but we know well enough in writing and conversation what it means for two inscriptions or utterances to be *equiform*. So we can make the following agreement.

Words are types   We assume that throughout any particular discussion equiform words have the same properties of interest to us for reasoning. We therefore identify them and treat them as the same word. Briefly, *a word is a type*.

Some say that types are abstract objects. We cannot point to a type, only to a representative of it, for example, "Ralph is a dog." But all that we use and need in our reasoning is a process of identification of utterances or inscriptions. The reason we make those identifications may be by our intuiting or having some non-sensory access to abstract objects called "types," or it may be due to our psychology, or physiology, or culture.

The assumption that words are types, in this sense, is an abstraction from experience, but it is also an agreement to limit ourselves. Though a useful abstraction, it rules out many sentences we can and do reason with quite well. For example, we shall have to distinguish the three equiform inscriptions in "Rose rose and picked a rose," using some device such as "$Rose_1$ $rose_2$ and picked a $rose_3$" or "$Rose_{name}$ $rose_{verb}$ and picked a $rose_{noun}$."

Now consider a simple piece of reasoning:

If Socrates was Athenian, then Socrates was Greek.
Socrates was Athenian. Therefore, ...

If claims are inscriptions, then the two equiform occurrences in this are distinct claims. How can we reason with them?

Since words are types, we can argue that the two equiform inscriptions here are both true or both false. It doesn't matter where they're placed on the paper, or who said them, or when they were uttered. Their properties for reasoning depend only on what words and punctuation appear in them in what order. Any property that

differentiates them isn't of concern to reasoning. We can make
the following agreement.

Claims are types In the course of any reasoning, we consider an
uttered or written piece of language to be a claim only if any other
piece of language that is written or uttered and that is composed of
the same words in the same order with the same punctuation can be
assumed to have the same properties of concern to our reasoning during
that analysis. We therefore identify equiform inscriptions or utterances
and treat them as the same claim. Briefly, *a claim is a type*.

Again, to say that we can make this agreement does not necessarily
mean it is just a convention. I do not deny that there might be good
reasons for our agreements, say, that there are abstract objects called
"types of sentences" which, perhaps without our being aware, have
led us to this agreement or at least justify the agreement as the only
possible way to proceed in reasoning.

But more fruitfully, we can see the agreement that we identify
distinct utterances or inscriptions as being the same for all our logical
purposes as an abstraction from experience, which is all we need when
we say that *Ralph is a dog* and RALPH IS A DOG are representatives of
the same type. Utterances are loud or soft, spoken clearly or mumbled.
Inscriptions are typed or handwritten, in blue ink or black or red; they
are on this page or that page or on a computer display. We are ignoring
certain aspects of our experience in order to simplify our reasoning.
This is part of the general procedure of creating a model of how to
reason well. It is important at every stage to note what abstracting
we do. If we encounter problems, contradictions, or counterintuitive
consequences of our work, we can go back and see if perhaps part of
what we chose to ignore in our experience might matter in a particular
context.

For example, consider:

box **R** → | The sentence in box **R** contains forty-seven letters. |

The sentence in box **R** contains forty-seven letters.

The two sentence inscriptions are equiform: they contain the same
words in the same order with the same punctuation. But they are

different for our reasoning. One refers to itself, one does not. To test the truth of the one on top, we need to consider whether it itself contains forty-seven letters, but that's not the case with the lower one. There is a clear and perceptible difference for our reasoning that is obscured by identifying them. In this case, though, the difference doesn't seem to matter.

But now consider:

box **S** → | The sentence in box **S** is not true. |

The sentence in box **S** is not true.

The inscription in the box is an example of the liar paradox, which seems to be both true and false. The inscription below it is not a version of the liar paradox but refers to the one above. The differences between these equiform inscriptions matter.[7]

The division of true and false

What is true is not false; what is false is not true. The world is, and our language, if used correctly, does not describe both what is and what is not with the same words. If "Ralph is a dog" is true, it can't be false; if "Ralph is a dog" is false, it can't describe how the world really is.

This, we believe, is the nature of truth because we believe that there is a world external to us, parts of which we are trying to describe. The world is coherent, out there; and if we use language correctly, no claim can be both true and false.

But then what about the apparent claims we make in ethics? What in the world is there that could make "You should never kill a dog" true or false? Yes, there are two categories for such sentences: good/bad, or correct/incorrect, or just/unjust, or assertible/unassertible, or . . . , yet those divisions are not dividing claims into the true and false. But compare these examples.

Example 3 Physician: Don't smoke anymore.
  Matilda: O.K.

  Analysis Suppose that Matilda then goes out and smokes a couple cigarettes. We'd say she is perverse, or stupid, or she just didn't follow the doctor's orders. There's no question of belief or truth.

Example 4 Physician: You shouldn't smoke anymore.
  Matilda: I agree.

Analysis Suppose again that Matilda goes out and smokes a couple cigarettes. In this case we think she can be charged with inconsistency (if she hasn't changed her mind). That's because Matilda's attitude about "You shouldn't smoke" is one of belief. The doctor is not commanding her; such a conversation would typically be preceded or followed by an attempt by the doctor to convince her that she shouldn't smoke. And belief is belief that something is true.[8]

Perhaps, though, what Matilda is asked to believe is that the prescriptive claim is good advice. But to say that "You should stop smoking" is good advice is just to say that you should stop smoking.

The word "true" is odd in that we get nothing new by ascribing it to a sentence. That's equally so for these other divisions. Compare:

"Dick is an American" is true.
Dick is an American.

"New York is in the United States" is assertible.
New York is in the United States.

"Matilda should stop smoking" is good advice.
Matilda should stop smoking.

It's only when surveying or analyzing the use of the notions of truth and falsity and constructing theories that the words "true" and "false" play a significant role. *The labels* is assertible *and* is good advice, *just like* is true, *are shorthand for the conditions we look for in evaluating whether to accept a claim.* If we have those conditions, we might as well assert the claim.[9] It may be a simplification, but if so it is a simplification of great utility to seize on the following similarities:

These are the conditions under which you are justified in believing the sentence ~~is true~~.

These are the conditions under which you are justified in believing the sentence ~~is assertible~~.

These are the conditions under which you are justified in believing the sentence ~~is good advice~~.

The resistance to viewing various divisions we use in our reasoning as divisions into the true and false follows, I suspect, from a particular metaphysics of truth and falsity: the true is what corresponds to the case, and that is independent of us and our interests. These other notions, it is said pejoratively, are dependent on our human capacities and interests, so they can't really be the division into the true and false. That is one particular metaphysics, a metaphysics that is part of a view of reasoning in which human capacities and interests are not considered constraints on models of good reasoning and the search for truth.

The divisions are dichotomies

The idea that what is true is not false, and that what is false is not true seems not only apt but obvious when we think of truth in terms of some kind of correspondence with a world external to us. But why should divisions into correct/incorrect, good/bad, ethical/unethical be mutually exclusive? Why can't we have a claim that is both ethical and unethical, or one that is good advice and not good advice, or one that is assertible and not assertible? The nature of our personal evaluations doesn't seem to rule that out.

Personal evaluations, at least of the type that give rise to a true/ false division, are implicitly if not explicitly prescriptive. If it is good advice, you should do it; if it is bad advice, you shouldn't do it. If it is correct, then you should do it that way; if it is incorrect, then you shouldn't do it that way. If it is good, you should approve of it; if it is bad, you should disapprove. If a claim such as "You should never torture a dog" is both good advice/ethical/correct and bad advice/ unethical/incorrect, then we are enjoined to both do and not do, to act and to act in a contrary manner. That we cannot do.

The *law of excluded middle* (every claim is true or false) and the *law of non-contradiction* (no claim is both true and false) are called "laws" by some because those principles lie at the heart of their meta-physics. But we can just as well see them as rules to simplify our reasoning. Or we can see them as reflecting a human capacity or need to classify as either-or. We adopt them as the basis of all our divisions into the true and false. We take each such division to be a dichotomy.

Even those who propose systems of reasoning based on the idea that there are degrees of truth, where a strict division is denied, impose a division: a line is drawn that says on this side are claims we can use to derive further claims on which we can rely, the ones with "desig-nated" values, the assertible ones, while on the other side are claims with undesignated values.[10] Between affirming and denying there seems to be no third choice.

But then what do we do in reasoning with a claim such as "You should never kill" when we find we have as good reason to believe it is true as to believe that it is false? We suspend judgment. But what if there is nothing more that could determine which it is: true or false?

We agree to view the sentence as a claim in order to determine whether it is true or whether it is false. But in doing so, we may find that the truth-conditions for it are indeterminate. We cannot then say that the sentence is both true and false without abandoning all our methods of reasoning, including those that led us to that conclusion. So we no longer agree to view the sentence as true or false but not both. We no longer take the sentence to be a claim.

Thus, physicists reasoned with "The ether has no mass" only to find that since there was no ether the sentence was neither true nor false, and so we no longer accept the sentence as a claim. We take the sentence "This sentence is false" to be a claim, and in reasoning with it find that if it is true then it is false, and if it is false then it is true. We reason with it on the assumption that it is a claim in order to find that it is neither true nor false, and then no longer accept it as a claim.

So let us revise our definition of "claim."

Claims A claim is a written or uttered piece of language that we agree to view as being either true or false but not both.

It is because we take these divisions to be dichotomies that we can devise methods of how to reason across a variety of subjects and purposes. These divisions function the same in our reasoning. The notions of possibility, inference, valid inference, and strong inference are all defined relative to the true/false division as a dichotomy, and all have been defined relative to both subjective as well as objective divisions.[11] I have shown how this is so for formal systems of reasoning, as I describe in "A General Framework for Semantics for Propositional Logics" in this volume. And in this series, *Essays on Logic as the Art of Reasoning Well*, I have used such dichotomies to establish rules for reasoning with arguments, explanations, mathematical proofs, reasoning in the sciences, conditionals, subjective claims, and prescriptive claims.

The false is what is not true

Given a dichotomy that functions as true and false in our reasoning, we have to ask whether certain utterances that appear to be claims but are stupid, or senseless, or bad, or worthless (depending on the particular notion we're investigating) should be classified as claims.

For example, "Green dreams jump peacefully" is a declarative sentence. But it doesn't make any sense. So without further qualification we have no motive to reason with it, and hence we don't take it to be a claim.

In contrast, when Suzy said to Dick, "You should hold your breath for four minutes in order to stop hiccuping," we do indeed want to reason with the sentence. It's stupid, silly, but more, it's bad advice or wrong, so we classify it as false.

When someone says to us "The King of France is bald," it seems to be nonsense since there's no King of France. Yet that sentence might show up in our reasoning. It's stupid and wrong, but we can treat it as a claim, a false claim, for it certainly isn't true.

The problem of whether to jettison, to disregard sentences that are odd, or nonsensical, or too vague arises in a quite general way when we employ a formal language as a guide to the formation of propositions with which we'll reason. We carefully distinguish between syntax and semantics. The description of the formal language invokes no semantic aspects of the primitive parts of the language: we set out how to make well-formed-formulas solely in terms of structure and parts of speech. This division of form and content is invoked, usually implicitly, for every formal logic.[12] If we do not divide form and content, there are immense problems in specifying what counts as a sentence of the formal language, and we cannot use the method of proof by induction on the form of formulas in analyzing the formal syntax and semantics.[13]

Once we have a formal language, we take certain linguistic expressions to realize the primitive symbols of the formal language, in the case of predicate logic both predicates and names (or linguistic expressions of those if you are a platonist). Then we have no choice, due to the division of form and content, but to say that every expression using these predicates and names that instantiates a formula of the formal language is meaningful and that every expression that instantiates a closed formula is a claim. That is, once we determine that the primitive parts of speech are meaningful, we commit ourselves to the entire stock of realizations of closed formulas as being not only meaningful but claims.

Suppose, then, that we take "is a dog" to be a predicate and the universe—the things we're reasoning about—to be all living animals. Suppose also that we take "Anubis" to be a name used to refer to a particular animal in that universe. And suppose that Anubis is a wolf-dog hybrid that eludes classification as either a dog or not a dog. Then "Anubis is a dog" is too vague to be taken as a proposition, yet we have agreed to accept it as one. In almost any realization of the formal language of predicate logic that we would wish to use as a guide to reasoning in our ordinary lives, there will be sentences that we know are too vague to have a truth-value but which we are committed to treating as claims.

We can deal with such sentences that we would normally classify as nonsense or too vague to be claims by classifying them as false for the convenience of having a clear analysis of form distinct from meaning. Thus, the formalization of "The King of France is bald" is classified as false in predicate logic, and its negation is true. Thus, the formalization of "If the moon is made of green cheese, then $2 + 2 = 4$" is classified as false in relatedness propositional logic, and its negation as true. Thus, "Anubis is a dog" is classified as false, and so its negation is classified as true.

It may be a simplification to call such sentences true or false rather than senseless or stupid, but if so, it is a simplification of considerable utility. So we classify them as claims. But not as true claims. *They are false because they are not suitable to proceed on for deriving true claims*. This is how people treat all the notions that yield dichotomies that we take to be true-false divisions: *falsity is the default truth-value*. A claim must pass certain tests in order to be true; all other claims are classified as false, whether those tests are for assertibility, or for good advice, or for being sufficiently probable, or . . . .[14]

The problem of accommodating sentences that are nonsense or are too vague to be taken as claims into our stock of propositions when we use a formal logic is due to our attempt to give clear, usable models of reasoning. It is not a problem that is inherent in how to reason well but only in how to adopt a guide for how to reason well. The solution of it does not depend on nor illuminate our understanding of what is nonsense or a vague sentence. The solution is simply part of the apparatus, part of how we abstract from our experience in order to give a clear model. Only to the extent that we take conditions for a claim to be true as primary and falsity as the default truth-value can such a solution be said to reflect our deeper commitments to our notion of truth.

When is a claim true?

We say that "Dick likes chocolate ice cream" was true a year ago but is false today. But if that's really what we mean, then we have serious problems in saying how to reason with claims that can be true at one time and false at another. It is better to recognize that we are implicitly indexing the sentence with the time about which it is meant, so that there are two claims under discussion: "Dick liked chocolate ice cream a year ago" and "Dick likes chocolate ice cream today," the first of which is true and the second false. In such cases we can always mark a claim with the time it is meant to describe.

But consider what Zoe said to Dick, "You should lose weight." If understood as meaning that losing weight will best achieve Dick's ends, then whether it is true or false is going to depend on Dick evaluating his various goals and how likely they are to be achieved or not achieved if he loses weight.[15] Dick's evaluations will change as he contemplates them, so it seems we can say that the claim is true or false only after Dick deliberates on it.

The problem of whether an utterance is a claim from the moment it is spoken or only after someone, or some group of people, or some creature makes an evaluation is endemic to all reasoning in which there is a subjective element. But it is a problem only if we adopt an attitude that there is an objective standard that a sentence must pass in order for it to be a claim: whatever conditions that determine whether it is true or false are satisfied or not at the moment the sentence is uttered. That standard seems not to be fulfilled in the claim about losing weight. But the adoption of that standard is also at issue.

Claims about the future have a similar problem. We all use and treat as a claim "It will rain tomorrow." Do we need to commit to the view that everything in the world is determined in advance for it to have a truth-value now?

The definition of "claim" is meant to avoid taking a stand on these issues. A sentence is a claim when we agree to view it as true or false. In all these cases it simplifies our reasoning enormously to view such sentences as true or false from the start of our consideration of them. If you wish, you can say that a proposition is what *is* true or false, and then what we are doing in our reasoning is establishing whether a particular sentence represents or correlates to a proposition. But then you're faced with explaining exactly what a proposition is, and how, if it is not the utterance in context, it can and does play a role in our reasoning. My sense is that invoking such propositions only serves to make it possible for those who want completely objective standards in their reasoning to have an object that is true or false, even when that object has no role other than marking a place we want to get to in our reasoning. More apt is to say that we reason with a sentence as if it is true or false in order to come to an evaluation of which it is, or to find that it is neither and hence not a claim.

Conclusion

We reason to arrive at truths. To do that we need to understand what we mean by saying that something is true or false.

Utterances or inscriptions that we call true or false are part of the experience of all of us in our reasoning. We can abstract from them to talk of types as true or false. In this way we can come up with a notion of claim that can be used in reasoning by people who take very different views of what kinds of things are true or false.

To say that an utterance is a claim is to agree to view it as true or false. That need not mean that it is true or false but only that we have some motive to treat it so, perhaps eventually finding that it is not suitable to be called true or false. That agreeing is often implicit, part of the general way we negotiate meaning.

The true/false division is often thought of as an objective classification of claims, corresponding in some way to how the world is. But there are many divisions we use in our reasoning that are based on what are apparently subjective evaluations. We can and do use divisions of utterances into good/bad, or acceptable/unacceptable, or assertible/unassertible, or right/wrong in the same way we use an objective notion of truth and falsity.

Those divisions are taken to be dichotomies: what is true is not false, and what is false is not true. Normally falsity is taken as the default truth-value so that we can accommodate odd or too-vague sentences into our reasoning.

In this way we have a general basis for reasoning that allows us to use the same methods across a wide variety of subjects and purposes.

Appendix 1   Are claims true or only represent what is true?

Claims are what we use in our reasoning. Some say, though, that what is true or false is not the utterance or inscription but the meaning or thought expressed by that, what they call a "proposition" or "mental proposition." So the following, if uttered at the same time and place, all express or stand for the same proposition:

(**)   It's raining.
    Il pleut.
    Pada deszcz.
    Está chovendo.

The word "true," they say, can be properly applied only to things that cannot be seen, heard, or touched. Sentences "express" or "represent" or "participate in" such propositions.

Platonists take this one step further. A *platonist*, as I use the term, is someone who believes that there are abstract objects not perceptible to our senses which exist independently of us. Such objects can be perceived by us only through our intellect. The independence and timeless existence of such objects, they say, account for objectivity in reasoning and mathematics. In particular, propositions are abstract objects, and a proposition is true or is false, though not both, independently of our even knowing of its existence.

But the platonist, as well as the person who thinks a proposition is the meaning of a sentence or is a thought, reasons with language. For me to reason with someone who takes propositions to be what is true or false, it is not necessary that I believe in abstract objects or thoughts or meanings. It is enough that we agree that certain utterances and inscriptions are or from his viewpoint represent propositions. Whether such a piece of language expresses a true proposition or a false proposition is as doubtful to him as whether, from my view, it is true or it is false. The question of whether the four inscriptions at (**) express the same proposition for him amounts on my view to whether we should identify those four inscriptions for all our purposes in reasoning.

From my perspective, the platonist conception of logic is an idealization and abstraction from experience. From the platonist perspective, I mistake the effect for the cause, the world of becoming for the reality of abstract objects. But we can and do reason together using claims, and to that extent the definition of claim presented in this essay can serve platonists or those who hold other views of propositions. In analyzing any particular kind of reasoning, we can take those views into account as added weight to the significance of what we take to be true or false.[16]

But platonists argue that taking claims as the basis of reasoning is hopeless. They say we cannot answer precisely the questions: What is a sentence? What constitutes a use of a sentence? When has a sentence been used assertively or even put forward for discussion? These, they say, can and should be avoided by taking things inflexible, rigid, timeless as propositions. But that only pushes back these problems to: How do we use logic? What is the relation of these theories of abstract objects to our arguments, discussions, and search for truth? How can we tell if this utterance expresses that abstract proposition? It's not that taking claims to be true or false brings up questions that can be avoided. The emphasis on precision and objective standards in reasoning has gone too far if we cannot relate our work to its intended use as a guide to reasoning well.

In contrast to platonists, some say that mental propositions, what might be called thoughts, are what are true or false. Alexander Broadie describes a medieval view of that and, from what I can gather, his own view in *Introduction to Medieval Logic*, p. 8:

> That we use the sounds or marks we do use in order to communicate is not a fact of our nature, for we could have used other signs, and other nations do use other signs. But what I think of when I think of what I call a "man" is the same as what a Frenchman thinks of when he thinks of an "homme", and as what a Greek thinks of when he thinks of an "anthropos". The thought is the same though the conventional express- ion of it differs. Thus the language of thought is universal in contrast to what we may term the "parochiality" of conventional languages. Indeed the intertranslatability of conventional languages is due precisely to the fact that, different as they are in respect of many of their characteristics, they can all be used to express the same set of thoughts.[17]

This is an example of the triumph of hope over experience. We all hope, desperately, to be understood by others. We do not want to think that we are so separate that no one can see the world as we do. We only need to make the effort, perhaps a great effort, to phrase our thoughts well and others will understand exactly what we are thinking. But every day in every conversation, we have evidence that others do not understand as we do, that the thought we wished to convey is not what the other person understood by what we said. Approximately sometimes, but exactly most certainly not. This is one of the most obvious and clearest conclusions we can make from our experience of "communicating" with others. Yet we persist in believing that others have the same thoughts as we do. Anyone who knows well two languages will, on reflection, admit that there is no real inter-translatability between them but only some approximation. Much is lost in translation: *traduire c'est trahir*. What is maintained is, roughly, something like the truth-conditions of sentences, and even that only approximately.

We are not sure what thoughts are, even for ourselves. We know that they needn't be linguistic (gestures as well as pictures in our minds). But what exactly they are, when we have had one, what distinguishes one from another, this we cannot say. How, then, can we proclaim that others have the same thoughts we do? We don't even know when we have the same thought we had an hour ago.

It is hard to see how the truth or falsity of our thoughts could be the basis for an analysis of how to reason well.

Appendix 2   Coherence rather than truth

Some say that we should abandon the view that what is true is what corresponds to the case, no matter how that might be interpreted. Claims are true or false, they say, as they cohere with other claims in our general understanding of the world.[18]

Coherence is invariably said to require consistency: the claim A does not cohere with B, C, D, ... if together they do not make a consistent collection of claims. But to invoke consistency is to invoke some methods of reasoning. On what basis should we accept those rules of reasoning? We would have no notion of validity, for that, too, depends on the notion of truth. We don't have any idea of what it means for a rule of inference to be acceptable if that isn't explained in terms of validity. We don't have any notion of what it means to take a claim as axiomatic, such as "Dogs are mammals or dogs are not mammals," unless we invoke the notion of tautology, which depends on a notion of truth. The whole enterprise of substituting coherence for truth is a sleight of hand, trying to divert our attention from the reliance on an informal, intuitive notion of truth to focus only on syntax. To say, for example, that "A or not A" is a fundamental principle that other principles have to cohere with, and we know this because, say, people use it that way, is to pretend that we can avoid all the hard work of trying to understand why we use that principle, what assumptions lie behind it, in favor of the superficial syntax. No reason is apparent even for why consistency should be demanded.

Appendix 3   Waismann on objective standards and the law of excluded middle

Many have written on the issues I discuss in this essay, too many to cite here. But one scholar, Friedrich Waismann, has discussed these issues in a way that particularly illuminates the analyses I have given.

In this first long extract from "Verifiability" (pp. 57–58), he discusses the interplay between convention and objective standards.

> Suppose there is a tribe whose members count "one, two, three, a few, many". Suppose a man of this tribe looking at a flock of birds said "A few birds" whereas I should say "Five birds",—is it the same fact for him as it is for me? If in such a case I pass to a language of a different structure, I can no longer describe "the same" fact, but only

another one more or less resembling the first. What, then, is the objective reality supposed to be described by language?

What rebels in us against such a suggestion is the feeling that the fact is there objectively no matter in which way we render it. I perceive something that exists and put it into words. From this it seems to follow that fact is something that exists independent of and prior to language; language merely serves the end of communication. What we are liable to overlook here is that the way we see a fact— i.e., what we emphasize and what we disregard—is *our* work. "The sun beams trembling on the floating tides" (Pope). Here a fact is something that emerges out from, and takes shape against a background. The background may be, e.g., my visual field; something that rouses my attention detaches itself from this field, is brought into focus and apprehended linguistically; that is what we call a fact. A fact is noticed; and by being noticed it becomes a fact. "Was it then no fact before you noticed it?" It was, if I *could* have noticed it. In a language in which there is only the number series "one, two, three, a few, many", a fact such as "There are five birds" is imperceptible.

To make my meaning still clearer consider a language in which description does not take the form of sentences. Examples of such a description would be supplied by a map, a picture language, a film, the musical notation. A map, for instance, should not be taken as a conjunction of single statements each of which describes a separate fact. For what, would you say, is the contour of a fact? Where does the one end and the other begin? If we think of such types of description, we are no longer tempted to say that a country, or a story told in a film, or a melody must consist in "facts". Here we begin to see how confusing the idea is according to which the world is a cluster of facts—just as if it were a sort of mosaic made up of little coloured stones. Reality is undivided. What we may have in mind is perhaps that *language* contains units, viz. *sentences*. In describing reality, describing it in the form of sentences, we draw, as it were, lines through it, limit a part and call what corresponds with such a sentence a fact. In other words, language is the knife with which we cut out facts. (This account is simplified as it doesn't take notice of *false* statements.) When we pass to a symbolism of language that admits of no sentences, we are no more inclined to speak of facts.

. . . Just as we have to interpret a face, so we have to interpret reality. The elements of such an interpretation, without our being aware of it, are already present in language—for instance, in such moulds as the notion of thinghood, of causality, of number, or again in the way we render colour, etc.

Noticing a fact may be likened to seeing a face in a cloud, or a
figure in an arrangement of dots, or suddenly becoming aware of the
solution of a picture puzzle: one views a complex of elements as one,
reads a sort of unity into it, etc. Language supplies us with a means
of comprehending and categorizing; and different languages categorize differently.

"But surely noticing a face in a cloud is not inventing it?"
Certainly not; only you might not have noticed it unless you had
already had the experience of human faces somewhere else. Does this
not throw a light on what constitutes the noticing of facts? I would not
dream for a moment of saying that I *invent* them; I might, however, be
unable to perceive them if I had not certain moulds of comprehension
ready at hand. These forms I borrow from language. Language, then,
contributes to the formation and participates in the constitution of a
fact; which, of course, does not mean that it *produces* the fact.

E.H. Gombrich in *Art and Illusion* fleshes out this view of "facts" in
exploring psychological bases of representation. In "The World as Process"
in *Essays on Language and the World*, I show a greater dependence of our
conception of "reality" on language than either of these authors do.

In this second long extract from "How I See Philosophy" (pp. 8–10),
Waismann discusses the law of excluded middle and sentences about the future.

This doubt has taken many different forms, one of which I shall single
out for discussion—the question, namely, whether the law of excluded
middle, when it refers to statements in the future tense, forces us into a
sort of logical Predestination. A typical argument is this. If it is true
now that I shall do a certain thing tomorrow, say, jump into the Thames,
then no matter how fiercely I resist, strike out with hands and feet like a
madman, when the day comes I cannot help jumping into the water;
whereas, if this prediction is false now, then whatever efforts I may
make, however many times I may nerve and brace myself, look down at
the water and say to myself, "One, two, three—", it is impossible for me
to spring. Yet that the prediction is either true or false is itself a
necessary truth, asserted by the law of excluded middle. From this the
startling consequence seems to follow that it is already now decided what
I shall do tomorrow, that indeed the entire future is somehow fixed,
logically preordained. Whatever I do and whichever way I decide, I am
merely moving along lines clearly marked in advance which lead me
towards my appointed lot. We are all, in fact, marionettes. If we are not
prepared to swallow *that*, then—and there is a glimmer of hope in the
"then"—there is an alternative open to us. We need only renounce the
law of excluded middle for statements of this kind, and with it the

validity of ordinary logic, and all will be well. Descriptions of what will happen are, at present, neither true nor false. (This sort of argument was actually propounded by Lukasiewicz in favour of a three-valued logic with "possible" as a third truth-value alongside "true" and "false".)

The way out is clear enough. The asker of the question has fallen into the error of so many philosophers: of giving an answer before stopping to ask the question. For is he clear what he is asking? He seems to suppose that a statement referring to an event in the future is at present undecided, neither true nor false, but that when the event happens the proposition enters into a sort of new state, that of being true. But how are we to figure the change from "undecided" to "true"? Is it sudden or gradual? At what moment does "it will rain tomorrow" begin to be true? When the first drop falls to the ground? And supposing that it will not rain, when will the statement begin to be false? Just at the end of the day, at 12 p.m. sharp? Supposing that the event *has* happened, then the statement *is* true, will it remain so for ever? If so, in what way? Does it remain uninterruptedly true, at every moment of day and night? Even if there were no one about to give it any thought? Or is it true only at the moments when it is being thought of? In that case, how long does it remain true? For the duration of the thought? We wouldn't know how to answer these questions; this is due not to any particular ignorance or stupidity on our part but to the fact that something has gone wrong with the words "true" and "false" applied here.

If I say, "It is true that I was in America", I am saying that I was in America and no more. That in uttering the words "It is true that—" I take responsibility upon myself is a different matter that does not concern the present argument. The point is that in making a statement prefaced by the words "It is true that" I do not *add* anything to the factual information I give you. *Saying* that something is true is not *making* it true: cf. the criminal lying in court, yet every time he is telling a lie protesting, his hand on his heart, that he is telling the truth.

What is characteristic of the use of the words "true" and "false" and what the pleader of logical determinism has failed to notice is this. "It is true" and "It is false", while they certainly have the force of asserting or denying, are not descriptive. Suppose that someone says, "It is true that the sun will rise tomorrow" all it means is that the sun will rise tomorrow: he is not regaling us with an extra-description of the trueness of what he says. But supposing that he were to say instead, "It is true *now* that the sun will rise tomorrow", this would boil down to something like "The sun will rise tomorrow now"; which is nonsense. To ask, as the puzzle-poser does, "Is it true or false *now* that such-and-such will happen in the future?" is not the sort of question to which an answer can be given: which *is* the answer.

Waismann uses the redundancy of the use of the phrase "is true" in the cases he is concerned with. I invoke that redundancy in a different way, to assimilate various classifications of claims to the single division into the true ones and the false ones. In contrast, I evade the issue of the import of the law of excluded middle with a definition of "claim" that allows for but does not suppose an objective basis for the classification into the true and false.

Notes

1. (p. 57) Some philosophers talk of "practical reason" in which no language is involved. It is not clear that practical reason has anything to do with reasoning in which the goal is to arrive at truths. See the section on dispositional rationality in the essay "Rationality" in *Prescriptive Reasoning*.

There are also some who point to experiments with birds and animals to say that reasoning does not require language. See, for example, *The Animal Mind* by James L. Gould and Carol Grant Gould, especially pp. 174–177. If that claim should prove true, then this essay and my other works should be seen as investigating reasoning as confined to language only.

2. (p. 58) Gottlob Frege in "The Thought: A Logical Inquiry" apparently takes this to be an acceptable inference.

3. (p. 58) Some say that a sentence such as "$2 + 2 = 4$" is fully precise and we understand it without any recourse to experience. Indeed, that sentence has been taken as both the archetype and standard for what we should accomplish in clarifying all our reasoning. But I show how our understanding of that sentence does depend on experience in "Mathematics as the Art of Abstraction" in *Reasoning in Mathematics and Science* in this series.

More generally in "Models and Theories" in *Reasoning in Science and Mathematics* I show how scientific laws are not claims but only schemes of claims until we attempt to apply them and in doing so say how the terms in them are meant to be understood. This is exactly how we resolve the ambiguity in example (‡): the sentences in it are not true but only true enough on application; the inference is a scheme, not a particular, until we apply it.

4. (p. 60) One colleague objected that some community might agree to view "All blue ideas are beautiful" as having a truth-value, yet that doesn't mean it is true or false. From our perspective those people either have a very different notion of truth than we do or they understand the words in that sentence differently than we do. So long as their notion of truth functions in the framework of reasoning I describe below, we have no reason *as logicians* to say that they have a faulty notion of truth. The situation would be the same as when a community adopts a "deviant" arithmetic, as I discuss in "Why Are There So Many Logics?" in this volume. We respect the reasoning of others to try to understand them, rather than dismissing the reasoning because of our inability to see immediately what they are doing.

5. (p. 60) Compare what John Lyons says in *Introduction to Theoretical Linguistics*:

> We have no direct evidence about the understanding of utterances, only about *misunderstanding*—when something "goes wrong" in the

process of communication. If, for instance, we say to someone *bring me the red book that is on the table upstairs* and he brings us a book of a different colour, or a box, or goes off downstairs in search of the book, or does something totally unexpected, we might reasonably say that he has "misunderstood" the whole or some part of the utterance (other explanations are of course possible). If he does what is expected (goes off in the right direction and comes back with the right book) we might say that he has correctly understood the utterance. . . . Normal communication rests upon the assumption that we all "understand" words in the same way; this assumption breaks down from time to time, but otherwise "understanding" is taken for granted. Whether we have or have not the same "concepts" in our "minds" when we are talking to one another is a question that cannot be answered otherwise than in terms of the "use" we make of words in utterances. It would probably be true, but rather pointless, to say that everyone "understands" a particular word in a slightly different way. Semantics is concerned with accounting for the degree of uniformity in the "use" of language which makes normal communication possible.   p. 411

6. (p. 60) Compare Isaiah Berlin in "Verification":

A proposition [is] any sentence which conveys to someone that something is or is not the case. And this seems on the whole to accord with common usage.   p. 16

In some of my earlier writings I said that a claim is a declarative sentence *used in such a way* that it is true or false, in order to avoid engaging in this discussion of the nature of agreements. However, the example "Raining" above shows that what is true or false need not be a sentence. Worse, a declarative sentence is usually defined to be one that is true or false.

7. (p. 64) A resolution of the liar paradox that depends on distinguishing these is due to John Buridan, as translated and explained by George Hughes in *John Buridan on Self-Reference*. See my "A Theory of Truth Based on a Medieval Solution to the Liar Paradox" (revised in Chapter XXII of *Classical Mathematical Logic*) for a modern formulation of that.

8. (p. 65) Alan R. White in *Truth*, Chapter 3.b, surveys arguments for and against the view that prescriptions are true or false and comes to a similar position:

We often think of moral pronouncements as something which can not merely be agreed or disagreed with, argued about, or contradicted, but also as being about what can be discovered, assumed or proved, believed, doubted or known; all of which characterize what can be true or false.   p. 61

9. (p. 65) Compare Michael Dummett, *Elements of Intuitionism* p. 371:

> It is evident that it is fundamental to the notion of an assertion that
> it be capable of being either correct or incorrect; and therefore, in
> so far as assertion is taken to be the primary mode of employment of
> sentences, it is fundamental to our whole understanding of language
> that sentences are capable of being true or false, where a sentence is
> true if an assertion could be correctly made by uttering it, and false if
> such an assertion would be incorrect.

10. (p. 66) See Chapter VII of my *Propositional Logics*.

Timothy Smiley in his "Comment on 'Does Many-Valued Logic Have Any Use?' by D. Scott" says:

> The way to defend [the method of designating truth-values] is to read
> 'true' for 'designated'. The method of defining logical consequence
> then needs no justification, for it now reads as saying that a proposi-
> tion follows from others if and only if it is true whenever they are all
> true. What does need explaining is how there can be more than two
> truth-values. The answer is that propositions can be classified in
> other ways than as true or untrue, and by combining such a classif-
> ication with the true/untrue one we in effect subdivide the true and
> untrue propositions into a larger number of types. For example,
> given any property ϕ of propositions, there are prima facie four
> possible types of proposition: true and ϕ, true and not ϕ, untrue and ϕ,
> untrue and not ϕ. If ϕ is unrelated to truth, like 'obscene' or 'having
> to do with geometry', all four types can exist and we get four truth-
> values, two being designated and two undesignated. If ϕ has any
> bearing on truth some of the types may be ruled out; e.g., if ϕ is
> (perhaps) 'about the future' or 'meaningless', the type 'true and ϕ '
> will be empty, leaving three truth-values of which just one is desig-
> nated. One cannot foretell how the connectives will behave with
> respect to this or that classification of propositions, but to the extent
> that the types of compound propositions turn out to be functions of
> the types of their constituents, so we shall get a many-valued logic.
>
> pp. 86–87

11. (p. 67) A discussion of subjective and objective standards is given in "Subjective Claims" by Fred Kroon, William S. Robinson, and me in *The Fundamentals of Argument Analysis* in this series of books.

In *Truth as One and Many*, Michael P. Lynch takes truth to be a structural notion, too. But he bases his work on what he calls "folk beliefs" about
the nature of truth that are far different from what I take as fundamental here.
Yet he cites no studies to show that those assumptions are commonly held.
Arne Naess in *"Truth" as Conceived by Those Who Are Not Professional*

Philosophers debunks all the talk that philosophers make about what the common notion of truth is, about what ordinary folks believe. Naess does a sociological experiment, questioning people about their views, and shows that there is not only no unanimity but a huge variety of views of truth all held by a more or less equally small percentage of the people he surveyed. Cory D. Wright in "On the Functionalization of Pluralist Approaches to Truth" summarizes and discusses current debates about whether truth is univocal or pluralist as well as claims by Lynch and others about the structural nature of truth.

12. (p. 68) See "A General Framework for Semantics for Propositional Logics", pp. 33–34 in this volume.

13. (p. 68) See "On the Error in Frege's Proof" in this volume for an example of this.

14. (p. 69) This is so for not only the formal logics studied in *Propositional Logics* but also for informal methods and rules for reasoning well that are analyzed in the other volumes in this series. It is also crucial in developing a formal method of reasoning with non-referring names and resolving the liar paradox, as I show in *Classical Mathematical Logic*.

　　Some logicians, either because they believe that there are true contradictions, or because they believe that there are some claims that are both true and false, or because they wish to model how to reason in the presence of (possible) contradictions without collapsing all deductions into triviality, propose what are called *paraconsistent logics*. In those, it seems, truth is taken as the default truth-value with tests set out for a claim to be false. See, for example, Chapter IX of *Propositional Logics*. But I know of no such logic that has clear enough semantics to merit consideration as proposing a view of truth and falsity as opposed to formal methods to justify a deductive system. If such clear semantics can be given, then a mirror image of the view here would be needed. In "Paraconsistent Logics with Simple Semantics," I show that we need not take truth as the default truth-value in order to have clear and well-motivated semantics for a logic in which not all claims follow from a contradiction.

15. (p. 70) See "Reasoning with Prescriptive Claims" in *Prescriptive Reasoning* in this series.

16. (p. 72) Whenever abstract propositions or thoughts or meanings are invoked as what are true or false, it is the sentences that "express" them that are pointed to as our access to them. For example, Gottlob Frege in "Negation" says:

　　How, indeed, could a thought be dissolved? How could the interconnexion of its parts be split up? The world of thoughts has a model

in the world of sentences, expressions, words, signs. To the structure
of the thought there corresponds the compounding of words into a
sentence; and here the order is in general not indifferent. To the
dissolution or destruction of the thought there must accordingly
correspond a tearing apart of the words, such as happens, e.g., if a
sentence written on paper is cut up with scissors, so that on each scrap
of paper there stands the expression for a part of a thought. p. 123

See also Frege's "The Thought: A Logical Inquiry."

Colwyn Williamson in "Propositions and Abstract Propositions" reviews
these and other views of what a proposition is from a viewpoint similar to mine.

17. (p. 73) Aristotle, *De Interpretatione*, 1, spoke similarly:

Spoken words are the symbols of mental experience and written
words are the symbols of spoken words. Just as all men have not
the same writing, so all men have not the same speech sounds, but
the mental experiences, which these directly symbolize, are the same
for all, as also are those things of which our experiences are the images.

18. (p. 74) See "The Coherence Theory of Truth" by James O. Young for a
survey and other criticisms of this view.

On Translations

An analysis is given of what counts as a translation between logics and of what counts as a translation that preserves meaning. The resulting criteria can then be used to evaluate formalizations of ordinary language propositions into formal logics, which in turn suggest criteria for evaluating translation between ordinary languages.

When we translate, we hope to capture all of the meaning of the original. That is our goal, but it cannot be our test for what counts as a good or even adequate translation, for we have no clear conception of meaning, not for words, not for phrases, not for sentences.

By considering translations between logics where the languages have clear structures and circumscribed meanings, we can give precise criteria for translations. Starting with the simplest kind of logics and working to more complex ones and then treating formalizing as a kind of translation, we can begin to see how to judge translations between ordinary languages.

Propositional logic

Propositional logics are formalizations of how to reason with propositions as wholes using abstractions of "and," "or," "it's not the case that," and "if . . . then . . .", for which we use the formal symbols ⌐, ∧ , ∨ , → .[1] We create a formal language using variables to stand for propositions. A logic is defined by giving models of the formal language: first propositions are assigned to the variables, and then semantic values from a restricted range specified in advance are given to those. A model is meant to capture all the meaning we are interested in for the propositions of the model. The truth-conditions of a claim such as "Ralph is a dog ∨ cats carry disease" are, for this logic, the models in which it is true.

The formalization of the notion of consequence is that a formula B is a consequence of a collection of formulas Γ if and only if in every model in which all the formulas of Γ are true, so too is A. When that is the case, we say that A is a *semantic consequence* of Γ or that the inference "Γ therefore A" is *valid* and write Γ ⊨A. Sometimes to emphasize that we are discussing a particular logic **L**, we write Γ ⊨$_\mathbf{L}$A. A *tautology* is a formula that is true in every model.

Sometimes a logic is specified syntactically. First, certain formal sentences are taken to be obviously true under all circumstances and are dubbed axioms. Then a formalization of proving is given, and the notion of consequence that it generates constitutes the logic. This syntactic characterization is usually a supplement to the definition of the logic in terms of models, in which case it is necessary to prove that the two notions coincide: what is a consequence in one is a consequence in the other. Though useful, an axiomatization is not essential in understanding what constitutes a good translation. If a logic is given only syntactically with no models explaining the semantic basis for the axioms or rules, we can have no clear idea why the particular axioms are chosen nor why that particular formalization of proving is considered right. We are left with no justification for the logic and no way to tell if a translation to or from it preserves meaning, for we can only point to formal consequences.

Many translations from one propositional logic to another have been given in the years since propositional logics were first clearly defined in the early part of the twentieth century.[2] To show that a mapping was a translation no general notion of logic or model was needed. This is a translation, it was said, and that assertion was or was not accepted by others. Now we want to make a precise definition of "translation" by which to judge those and others.[3]

Suppose, then, that we have two propositional logics:

L, with language L$_\mathbf{L}$ and a collection of models that define truth and consequence for that logic.

M, with language L$_\mathbf{M}$ and a collection of models that define truth and consequence for that logic.

Given a way $*$ to transform formulas of L_L to formulas of L_M, we'll notate the transformation of a formula A as A$*$, and the collection of transformations of formulas in Γ as $\Gamma*$.

We want that translations preserve meaning. But to start, let's consider meaning only as it is codified in the syntax of these logics.

Certainly truths should be mapped to truths, but the only truths we can point to when looking at just the syntax are the tautologies. So a translation should preserve tautologies. That is, given a mapping $*$ of L_L to L_M, we want that if $\vDash_L A$ then $\vDash_M A*$. But more, we want that if $\vDash_M A*$, then $\vDash_L A$, for translating a falsehood to a truth, much less a claim that can be false to one that must be true, is not acceptable as a formalization of the notion of an adequate translation. So for a translation, we require that $\vDash_L A$ iff $\vDash_M A*$.

This requirement, however, is not enough. Given any two propositional logics, there is a mapping of the tautologies of one onto the tautologies of the other, since both collections are countable. If each logic is either decidable or axiomatized, there is even a constructive mapping.

We can grasp more of the meanings of sentences via the syntax. The meaning of a sentence is its truth-conditions, and those are what figure in evaluating consequences. So if $\Gamma \vDash_L A$ then we should have that $\Gamma* \vDash_M A*$. And we expect further that the mapping introduces no new consequences: if $\Gamma* \vDash_M A*$, then $\Gamma \vDash_L A$. If this condition holds, tautologies are preserved by the mapping, too, since a tautology is a formula that follows from the empty collection of formulas.

Translations   Given logics **L** and **M**, a mapping $*$ from L_L to L_M is a *translation* iff for every collection of formulas Γ and every formula A, $\Gamma \vDash_L A$ iff $\Gamma* \vDash_M A*$.

This is our minimal notion of a translation.[4] But for the map to be a good translation, we also expect it to be a regular way to transform formulas, not simply *ad hoc*. Most mappings that have been offered as translations are indeed translations by this definition and have a strong grammatical regularity, satisfying the following.

Grammatical mappings   A mapping $*$ of a formal language is *grammatical* if and only if there are functions λ, φ, ψ, θ, and ρ such that for all atomic formulas p and formulas A and B in the language:

$$p^* \;=\; \lambda(p) \qquad (A \wedge B)^* \;=\; \theta(A^*, B^*)$$

$$(\neg A)^* \;=\; \varphi(A^*) \qquad (A \vee B)^* \;=\; \rho(A^*, B^*)$$

$$(A \rightarrow B)^* \;=\; \psi(A^*, B^*)$$

There are, nonetheless, translations that are regular and easy to specify that are not grammatical, such as the translation of classical propositional logic to intuitionistic propositional logic via $A^* = \neg\neg A$.

So far we can only point to the syntax of the languages and claim that the mapping somehow respects meaning because it preserves consequences. But in many cases it is possible to show that a translation respects the classes of models that define the logics and, hence, preserves meaning to the extent that it is captured in the models we have. Indeed, it is by this method that most of the translations in the past were shown to preserve consequences.

How can we use models to justify that a mapping is a translation from L_L to L_M? We need to show:

(1) If $\Gamma \vDash_L A$, then $\Gamma^* \vDash_M A^*$.

(2) If $\Gamma^* \vDash_M A^*$, then $\Gamma \vDash_L A$.

We can prove (1) by showing that if $\Gamma^* \nvDash_M A^*$, then $\Gamma \nvDash_L A$. If $\Gamma^* \nvDash_M A^*$, then there is a model M of M such that $M \vDash_M \Gamma^*$ and $M \nvDash_M A^*$. So we find a model M^* of L such that for all B:

(3) $M^* \vDash_L B$ iff $M \vDash_M B^*$

Then $M^* \vDash_L \Gamma$ and $M \nvDash_L A$. That is, we want to set up a correspondence of models as in this diagram:

(4)

If every model of L is M^* for some model M of M, then we can prove (2) in the same manner: If $\Gamma \nvDash_L A$, then there is some model L of L such that $L \vDash_L \Gamma$ but $L \nvDash_L A$; so there is some M such that $L = M^*$, and then $M \vDash_L \Gamma^*$ but $M \nvDash_L A^*$. Actually, we need only that for any model L of L there is some model M of M where M^* is *elementary equivalent to* L, that is, the same wffs are true in each. In that case we say that the mapping of models is *onto up to elementary equivalence*. We need not require that the class of models of L has been pared down so that for every pair of models there is at least one formula that distinguishes them: true in one and false in the other.

If we have (4), then the model M restricted to Wffs* = {A* ∈ L_M: A is a wff of L_L} determines the model M* of L via (3). The model M*, however, need not determine M.

This description is only suggestive of how to proceed because each logic has its own peculiar kind of models: two-valued for classical propositional logic, possible-worlds for intuitionistic propositional logic, three-valued for Łukasiewicz's propositional logic, and many others. So long as we use only the class of models of the originators of the logics, there doesn't seem to be more we can say about the nature of how translations do or do not respect meanings.

In order to give general rules for what counts as a translation that preserves meaning, we need an overview, some general formulation of what counts as a model for a propositional logic. Further, if we wish to show that there is no translation from a particular logic to another that preserves meaning—that one formalization of how to reason well is too narrow in its scope to account for reasoning well by another standard— we need a general notion of propositional logics and their models.[5]

Despite the wide variety of ways of presenting logics and models, we can give a uniform notion of models for propositional logics. As described in "A General Framework for Semantics for Propositional Logics" in this volume, in order to give a model we first specify a *semi-formal language*, which is the formal language with the propositional variables replaced by sentences of an ordinary language, sentences that we take to be (representatives of what is) true or false; these are the atomic propositions of the semi-formal language. We give meaning to the formal connectives and the formulas of the formal language by first assigning semantic values to the atomic propositions. Those values include whether each particular atomic proposition is true or false, and, depending on the logic, another value, some other content, such as subject matter or ways in which the proposition could be known to be true. Then we extend those values to all the formulas of the semi-formal language in a manner that is characteristic of that particular logic reflecting the understanding of the connectives but that is uniform across all logics varying only in how the additional content is handled.

The assignment of semantic values to the atomic propositions and the inductive procedure for extending those to all formulas of the semi-formal language constitute a *model*. In this way we are justified in saying that we have given meaning to the connectives and that every formula of the semi-formal language is a proposition.

For a particular logic, models given in terms of the uniform semantics are often derived from the models given by the originators of the logic. But in each case the semantics given within the general framework can be seen as accurately reflecting the motivations of the logics. The overall framework for the formal language and semantics can be understood as defining the study of propositional logics.

So suppose we have logics **L** and for **M** with classes of models for each of them in the form of the uniform semantics for propositional logics, and we have a mapping * from the language of **L** to the language of **M**. Suppose further that there is a mapping of models that satisfies diagram (4) and that recreates a model of **L** within a model of **M** so that **M*** is just **M** restricted to Wffs*: the truth-value of A in **M*** is the truth-value of A* in **M**, and the content of A in **M*** is the content of A* in **M**. Then if the mapping of models is also onto up to elementary equivalence and is grammatical, we say it is *semantically faithful* with respect to those classes of models.[6]

A semantically faithful translation satisfies all we can expect of a translation to preserve meaning within the restricted context of a formal language and formal models. Certainly it satisfies all we can expect given the restricted notions of meaning that are given for the logics. If a translation that is not semantically faithful is nonetheless claimed to preserve meaning, a different kind of analysis will be needed to show that. The only examples that have been proposed relax the condition on regularity of the translation so that it need not be grammatical.[7]

Many of the translations in the literature are semantically faithful.[8] For example, the usual translation of intuitionistic propositional logic to the modal logic **S4** is semantically faithful. Intuitionists who say that this translation does not accurately reflect their understanding of intuitionistic propositional logic have to argue that the semantics for intuitionistic logic, given in terms of possible worlds, from which the semantics in the uniform framework are derived, do not accurately reflect their intentions and understandings of the connectives.

On the other hand, all translations that have been given of classical propositional logic into intuitionistic propositional logic are not semantically faithful, which justifies the intuitionists' claim that such translations do not preserve meaning; they continue to find classical propositional logic incoherent.

Many examples can be found in the literature where one logic is said to be the same as another. Often that is because one is just a truncated form of the other in which some of the connectives can be defined in terms of the remaining ones. But no criteria are given for what is meant by "the same logic." One possibility now is to say that two logics are the same if there are semantically faithful translations in both directions. For example, consider dependence logic which takes the content of a proposition to be its referential content: how and what it refers to in some broad sense. Then for $A \to B$ to be true, not only must the truth-values of A and B satisfy the classical table for $\to$, the content of A must include that of B, modeling the notion that for a conditional to be true its consequent cannot extend beyond what's contained in the antecedent; the other connectives are evaluated as in classical logic. Dual dependence logic is the same except that for $A \to B$ to be true, the content of A must be contained in that of B. For these logics, the mapping that translates every atomic proposition to itself and every connective as itself except that $(A \to B)^* = \neg B^* \to \neg A^*$ establishes a semantically faithful translation in both directions. Now we can ask whether we should classify these as the same logic.

Predicate logic

The standard way to extend propositional logic in order to formalize more of ordinary reasoning is to consider the internal structure of atomic propositions in terms of the predicates and names of which they are composed. A predicate is (or is represented by) a proposition with all the names and pronouns replaced with blanks. Names are understood to be names of things, so that predicate logic is based on the metaphysics that the world is made up of things and that propositions are, at least to the extent to be studied in this logic, about things.

The ideas about translations between propositional logics generalize to predicate logics, where primitive content is ascribed not to atomic propositions as wholes but to predicates and to names. Though not many translations have been exhibited, they have the same character as their propositional counterparts.[9]

There are also translations from one theory to another within a particular predicate logic. The archetype is the translation in classical predicate logic from plane geometry, as axiomatized with the syntactic primitives of points, a betweenness relation, and a congruence relation on lengths between points, to the arithmetic of the real numbers, as axiomatized with the syntactic primitives $<, +, 0, 1$. Though these two theories have apparently different subject matter, we can translate in either direction with a model-preserving translation. That we can do so raises the question whether or in what sense they are the same theory.

Lesław Sczerba has given the general form of semantic transformations between theories in classical predicate logic,[10] and in *Classical Mathematical Logic* I relate that to the syntax of translations. There I also give a survey of translations within classical predicate and a general theory of translations within classical predicate logic, where the translation between plane geometry and the arithmetic of real numbers is presented. Every model-preserving translation within classical predicate logic I know of is composed of translations that: (i) define atomic predicates, (ii) establish an equivalence relation on the universe, (iii) relativize quantifiers (pick out a subuniverse), (iv) eliminate or add parameters, and/or (v) convert functions into predicates.

Second-order predicate logic

Second-order predicate logic extends the range of propositions that can be formalized in predicate logic. It allows quantification over not only things but also predicates, as in "Marilyn Monroe had all the qualities of a great actress." Given a semi-formal language, if a model of it can have a universe of predicates for quantification that need not be the collection of all predicates for that semi-formal language, then such quantification amounts to quantification over things of a special kind and can be "reduced" to first-order classical predicate logic. That reduction is a syntactic transformation that we can evaluate as a translation: it is regular and preserves consequences. Whether it is model-preserving depends on what we mean by "predicate": if predicates are mathematical abstractions, then the translations preserve meaning; if predicates are linguistic, then the translation is not model-preserving. The choice of metaphysical basis for second-order logic affects what we consider to be the meaning of propositions formalized in it.[11]

Formalization as translation

When we wish to formalize an argument in predicate logic, we transform its ordinary language propositions into ones in a semi-formal language, that is, a language in which the linguistic predicates (or representatives of them) and names that occur in the original proposition are used along with the formal propositional connectives and formal quantifiers and variables. For example, the following is a semi-formal wff:

$\forall x\,((x$ is a dog $\land \neg\,($Ralph barks$)) \to$
$\qquad (\neg\,($Ralph is a dog$) \lor \exists x\,(x$ is a dog $\land \neg\,(x$ barks$)))$

The semi-formal language inherits the structure of the formal language, as all propositions that can be formed in it are realizations of formal wffs.

We can view formalizations of ordinary language reasoning as translations from ordinary language to a semi-formal language. Ordinary language, in this case English, does not have a formal structure, or at least no one has been able to characterize that. But we can nonetheless use our work in translations between formal logics as a guide to how to analyze formalizations.

First, for a formalization to be a translation, consequences must be preserved. That is, since "Ralph is a dog and all dogs bark, therefore Ralph barks" is valid, its formalization must be a formally valid inference. But again, it is not just the preservation of valid consequences that is the goal of a translation; it is the preservation of truth-conditions. If "All dogs bark" is formalized well, then that proposition and its formalizations will have the same truth-conditions. If we preserve the truth-conditions for the formalizations of each of the ordinary language propositions we study, it will follow that consequences are preserved.

The difficulty is that we have no full analysis of truth-conditions for ordinary language propositions. Formal logics are meant to capture only part of what we intend as the meaning of such propositions, paying attention to only certain aspects of our speech and reasoning. Such restrictions are what make it possible to have a formal logic.

Thus, when we say that the formalization of an ordinary proposition into a formal logic preserves truth-conditions, we mean it does so relative to what we are paying attention to in the semantics of the formal logic. We cannot hope to have a fully meaning-preserving translation into a formal logic unless we say that the semantics of our formal logic captures all that is significant about the meaning of ordinary language propositions. Though that is sometimes said or implied about classical predicate logic, it is always meant as relative to the reasoning we are doing.

Given this background, we can set out some standards for a formalization into a predicate logic to be good. These standards, which I summarize here, are developed with many examples in *Predicate Logic*.

1. The formalization respects the assumptions that govern our choice of primitive syntactic categories of predicate logic and definition of truth in a model of the formal logic. The formalization also respects the assumptions that govern our choice of a particular logic.

Thus, we cannot formalize "Snow is white; this is snow; therefore this is white" in any predicate logic because "snow" cannot be understood as a name or a predicate: if it were a name, it would have to name some thing, but snow is not a thing, it is a mass; if it were a predicate, then "x is snow" would be a semi-formal wff but would be false for any reference of x since no thing is snow. Nor can we formalize in classical predicate logic "If the moon is made of green cheese, then $2 + 2 = 4$" if we consider the subject matter of propositions significant in our reasoning, for classical predicate logic does not take subject matter into account.

2. The formalization and the original would both be true or both be false relative to any given way the world could be. Moreover, the formalization and the original have the same additional semantic values (content) in any model of the logic we choose.

Further, if one proposition we are formalizing follows informally from another or a collection of other ordinary language propositions, the formalization of it is a formal semantic consequence of the formalization of those others.

In order to distinguish formalization from analysis and to try to ensure that the formalization will have the same semantic values as the original, we have an additional condition.

3. The formalization contains the same non-logical words as the original, allowing for changes of grammar to satisfy the first criterion.

Thus, the formalization of "Ralph is a bachelor" as "Ralph is a man $\wedge \neg$(Ralph is married)" is not good. But we do allow for the formalization of "All dogs bark" as "$\forall x$ (x is a dog $\rightarrow$ x barks)" where "dogs" is converted into "is a dog" as part of a general convention for formalizing plural common nouns.

These criteria for a formalization to be good are supplemented by others concerned with respecting tense and grammar as well as a parity of form requirement: similar ordinary language constructions go to similar formal language constructions, all else being equal. These are guidelines, which as more examples are studied can be refined and supplemented by more specific rules.

We can also ask about translating in the other direction, from a semi-formal language to ordinary English. For example, we can read "$\forall x$ (x is a dog $\rightarrow$ x barks)" as "For anything, if it is a dog, then it barks." Doing this might not take us back to the original proposition we intended to formalize with that semi-formal proposition, namely "All dogs bark," but that is acceptable since we consider those two ordinary language propositions to be equivalent relative to the assumptions of, for example, classical predicate logic. However, following any rules we devise, we will quickly obtain sentences in ordinary language that, though they can indeed be taken to be propositions, sound so odd that we hesitate to use them, as when we try to translate a semi-formal proposition that begins "$\forall x_1 \; \exists x_2 \; \forall x_3 \; \exists x_4 \ldots$". It is the difficulty of reasoning with complicated ordinary language propositions that leads us to adopt formal languages to guide us.

Translations from one ordinary language to another

When translating from, say, Portuguese to English our goal is for the English translation to have the same meaning as the Portuguese. However, we have no clear idea and certainly no formal model of the meaning of Portuguese or English sentences. By restricting our attention to these languages as used in reasoning, we can have a more accessible goal: preserving the truth-conditions of the original in the translation.

That, too, is difficult to make precise. We can substitute for that large goal the more modest one that consequences be preserved. If a proposition in Portuguese follows from one or many other propositions in Portuguese, the translation of that one must follow from the translations of the others. In our formal analyses we study valid inferences as the formalization of the notion of "follows from." But in ordinary language reasoning we have a more ample notion of "follows from" in which an inference need not be valid but can be just *strong*: it is not impossible but is unlikely for the premises to be true and conclusion false at the same time and in the same way. There are problems with using the notion of a strong inference in our reasoning, and translations between formal logics will be no guide to us in investigating those, as no formal logic has been devised to model that notion.[12] Hence, what guide we have in analyzing translations between ordinary languages will at best be for preserving the aspects of the languages that are significant for reasoning with valid inferences.

Given these restrictions, we can require some kind of regularity in our translations. We can say that the same words go to the same words in each proposition; for example, "gato" in Portuguese is always translated as "cat" in English, though we may specify certain exceptions in advance, as when in English we use "cat" as a pejorative. Also, the same grammatical construction is always transformed into a particular grammatical construction in the other language, again with some exceptions specified in advance. But more than this we are hard-pressed to say.

There is another possibility for bringing more regularity and clearer criteria to such translations. We can first translate from Portuguese, into, say, a semi-formal language of classical predicate logic. We have relatively clear criteria whether such formalizations are good, and we are clear about what aspects of reasoning and language we are paying attention to in our analysis of classical predicate logic. Then we can translate from that semi-formal language to one in which English words replace the Portuguese (linguistic) predicates and names according to our previously chosen dictionary. Thus, "$\forall x$ (x é um cachorro $\rightarrow$ x late)" is translated as "$\forall x$ (x is a dog $\rightarrow$ x barks)." Then we translate from that semi-formal language to ordinary English according to the rules for such translations that we already have.

A great difficulty, though, is specifying that dictionary. When do we translate, for example, "tenho saudades" as "I long for," when as "I miss," when as "I am homesick"? People who know both languages seem to be able to make such choices regularly and well, and it is not unreasonable to hope that such a dictionary could be made.

But before we can use classical predicate logic or any other predicate logic as a bridge in translating from one ordinary language to another, we need that the metaphysical assumptions of predicate logic are compatible with those ordinary languages. When and how we might determine that are difficult issues discussed in the last chapter of *Predicate Logic*.

Conclusion

We began by looking at the simplest formal languages with the simplest structural analysis of meanings: propositional logics. We said that a mapping must preserve consequences in order to be a translation. A grammatical translation further respects the structures of the languages, taking grammatical units to grammatical units.

When the logics come with a formal analysis of the structure of meanings associated with the formal languages, we can ask whether a translation preserves meaning. Sometimes we can show that directly with respect to the standard models of the logics. To show that there is no translation that preserves meaning, however, we need a uniform analysis of how we can assign meanings in logics. Then we can give general criteria for when a translation preserves meaning.

These methods and criteria can then be extended to more complex formal languages and structures of meaning, such as predicate logic.

Many of the ideas and criteria from analyses of translations between logics can be applied to formalizations of ordinary reasoning in a formal logic, judging formalizations as translations. Though not as precise as for formal logics, the criteria for judging what counts as a good formalization are clear enough to be a useful guide.

We can extend these analyses further to translations between reasoning in ordinary languages by considering formalizations and translations into and out of ordinary languages and formal logics. Though only a fragment of what we might hope to analyze in judging ordinary language translations, it is a start.

Notes

1. (p. 84) What follows can be extended to languages with additional formal connectives. See "Valid Inferences" in this volume for a fuller presentation of the formalisms discussed in this paper.

2. (p. 85) See Chapter X of my *Propositional Logics* for a survey.

For surveys to more recent times see "New Dimensions on Translations between Logics" by Walter A. Carnielli, Marcelo E. Coniglio, and Itala M. L. D'Ottaviano and "Translations between Logical Systems: A Manifesto" by Walter A. Carnielli and Itala M. L. D'Ottaviano. For a general study of translations between many-valued logics see "Society Semantics and Multiple-Valued Logics" by Walter A. Carnielli and Mamede Lima-Marques.

3. (p. 85) For the full theory of translations between propositional logics discussed in this paper, see *Propositional Logics*. W. Haas in "The Theory of Translation" discusses the issue of what can be preserved in translating along the lines that I discuss here.

The papers listed in the previous footnote are part of a new approach to translations that is more abstract than what is discussed here. See "What is a Logic Translation?" by Till Mossakowski, Razvan Diaconescu, and Andrzej Tarlecki for an overview of that approach.

4. (p. 86) Some logicians require only the "if" part of the definition for a mapping to qualify as a translation; see the papers in the previous two notes. Though such mappings may be of interest in tracking metalogical properties of specific logics, they cannot be said to be a formalization of the notion of translation.

5. (p. 88) Compare: To show that a function is computable, it is not necessary to have a general theory of computable functions, but to show that a function is not computable, it is necessary to have such a characterization.

6. (p. 89) See Chapter X of *Propositional Logics* for the details.

7. (p. 89) See Chapter X of *Propositional Logics*.

8. (p. 89) See *Propositional Logics* for examples.

9. (p. 90) See my "Relatedness Predicate Logic" with Stanisław Krajewski for an example.

10. (p. 91) See Szczerba's "Interpretability of Elementary Theories."

11. (p. 91) See *Predicate Logic* for second-order predicate logics generally and *Classical Mathematical Logic* for classical second-order predicate logic.

12. (p. 95) See the essays "Generalizing" and "Probabilities" in *The Fundamentals of Argument Analysis* in this series.

Reflections on Temporal
and Modal Logic

The most popular method of incorporating time into a formal logic
is based on the work of Arthur Prior. It treats tenses as operators on
sentences. In this essay I show a problem with that approach, a confusion of scheme versus proposition, which makes any system built in
that way incoherent. I will compare how other formal logics deal with
the scheme versus proposition distinction and find that only for formal
modal logics does the same problem arise. I then compare Prior's
approach to other ways of taking time into account in formal logics.

Introduction

There are three ways we take account of time in our reasoning. We
relate sentences with temporal connectives like "before," "after," and
"at the same time as." We use time-markers, indicators of specific
times such as "June 6, 1970," or "1983," along with words that pick
out unnamed times, such as "sometime" or "always." And we use
tenses, which are markers for relative times. In some languages the
latter are attached to the verb or predicate; in other languages a marker
is attached to a sentence, a paragraph, or even an entire story.

The most popular method now of incorporating time into a formal
logic is based on the work of Arthur Prior. It treats tenses as operators
on sentences. In this essay I will set out a problem about how such
systems treat sentences as schemes or as propositions. I will compare
how other formal logics deal with the same issue. Then I'll compare

Prior's approach to other formal ways of taking time into account in logical systems.

Prior's approach to temporal logic

Consider the sentence:

(1)   John loves Mary.

This is not a proposition. That is, it is not true or false.[1] In order to be true or false, references must be supplied for the words "John" and "Mary." But more, we need to know what time (1) is meant to describe. If "John" means the person John Paul Jones, and "Mary" means Mary Magdalene, and the time is meant to be November 16, 2012, then (1) is false. If "John" means the man who married Mary Schwartz Rodrigues of Socorro in 2010, and "Mary" refers to Mary Schwartz Rodrigues, then it is a true proposition about January 16, 2011, but is a false one if it is meant to be about April 13, 2008.

Even with fixed references for the names, then, we cannot take (1) to be a proposition unless a time is specified. Otherwise (1) could be true at some times and false at others. Yet a proposition is true or false, not both true and false, nor true sometimes and false another. When we encounter a sentence that appears to be true at some time and false at another, we know we have an ambiguity. We can resolve that ambiguity by specifying a time that the sentence is meant to describe. Thus, we have to view (1) as a *scheme* awaiting references for the names and a designated time in order to become a proposition. Let's assume in what follows that references have been supplied for "John" and "Mary."

To take account of relative time in reasoning, Arthur Prior suggested that we treat "in the past" and "in the future" as adverbs of sentences as wholes. In his approach there are two "temporal operators":

P   meant to be understood as "in the past"
    or "in the past it is true that"

F   meant to be understood as "in the future"
    or "in the future it is true that"

Those readings are ambiguous between "sometime in the past" and "always in the past" and between "sometime in the future" and "always in the future." Normally, "sometime in the past" and "sometime in the future" are what are meant in the formal systems.

Using these operators, we can convert (1) into two distinct propositions:

P (John loves Mary)
F (John loves Mary)

Each is true or false. Taking the default time indication to be the present, such a system treats (1) as being about now, the ambiguity between that and the scheme being resolved by context.

Let's consider:

(2) P (John loves Mary)

This is meant to be understood and analyzed as:

(3) At some time in the past, "John loves Mary" is true.

In this methodology, "P" is said to be a operator on propositions. But the sentence (1) as it appears in (2) cannot be a proposition. If (1) were a proposition within (2), it would have to be about a time, and prefacing it with "in the past" would make no sense. For example, if (1) in (2) were taken to be a proposition about now as I'm writing, February 11, 2013, then (3) would mean:

In the past (John loves Mary on February 11, 2013)

And that's just nonsense.

Rather, P takes (1) as a scheme and converts that into the proposition (2), where we understand the past as relative to the time I am writing this. That's why I've used quotation marks around (1) in (3).

The sentence (3) gives the conditions under which (2) is true. It says that there is some time in the past that could be assigned to (1) that makes it into a true proposition. The formal semantics then build on that. I'll describe those briefly.

First, some conception of time is given. Let's assume for this discussion that time is linear and that it is made up of points, whether those be tiny, such as a nanosecond, or large, like last week. So long as those are linearly ordered, we have a timeline T. Implicit in (3) is that at each time, each atomic scheme, such as (1), is either true or false. That's formalized by assigning to each atomic scheme p a subset of the timeline: those times at which the scheme is true when interpreted as being about that time. Let's note the assignment of times to atomic schemes by υ. Further, we need to take some time n as being the now of our timeline. Then in the formal semantics:

P (John loves Mary) iff there is some t in T such that
    t is before n and t is in v (John loves Mary)

For "F (John loves Mary)" the condition is the same except "after" replaces "before."

Already a problem needs to be resolved. Suppose that "January 6, 1447" is a time in T. At that time, the people to whom "John" and "Mary" refer did not exist. Some would say that "John loves Mary" evaluated at that time is then nonsense. But that is not allowed here. On this approach, any atomic sentence that is evaluated at a time at which one or more names lacks reference must be treated as false. Falsity is the default truth-value.

We can extend these semantics to make each time a model not just of the atomic schemes but of combinations of those using the formal connectives ∧ for "and" and ¬ for "not," using, for example, the classical interpretation of those connectives. Thus, for every time t in the timeline we will have a model M_t of classical propositional logic. The collection of those models constitutes the model M.

This is clear enough. But in this approach the operators P and F can be iterated, as in:

P P (John loves Mary)

P F (John loves Mary)

F P (John loves Mary)

F F (John loves Mary)

P P P (John loves Mary)

P F P (John loves Mary)

Consider, for example,

(4) F P (John loves Mary)

Its truth conditions should be:

At some time in the future, "P (John loves Mary)" is true.

But "P (John loves Mary)" is a proposition, evaluated by (3). It is either true or false, and that can't change by prefacing it with F.

The usual presentation of the semantics for such a system obscures this issue. The truth-conditions for (4) are said to be:

(5) At some time in the future, at some time in the past relative to that time, "John loves Mary" is true.

Using $\models$ to stand for "true in the model," condition (5) is:

There is a time $t > n$ such $M_t \models P$ (John loves Mary)

which is iff there is a time $t > n$ such that there is a time $t' < t$ such that $M_{t'} \models$ John loves Mary.

The P in (4) is no longer meant to be relative to now as in (3) but to a time in the future.

Here again we can see that the operators "P" and "F" are not propositional operators. They convert propositional schemes into propositions. But that means that we have to view "P (John loves Mary)" as a proposition in (2) and as a scheme in (4). And similarly, we have to view the compound "$\neg$ (John loves Mary)" as a proposition yet as a scheme in "P ($\neg$ (John loves Mary))." It seems that we have an endemic ambiguity of scheme vs. proposition in this formal logic. Does the same ambiguity occur throughout formal logics?

Classical propositional logic

In classical propositional logic, we start with a few "connectives" from our ordinary language and formalize those. Typically we start with "and," "or," "not," and "if . . . then . . .". For our purposes here it's enough to consider just "and" and "not."

Let's suppose, then, that we are reasoning with these sentences:

John is a man.
Mary is a woman.
John loves Mary.

Let's assume that the references of "John" and "Mary" are fixed. And we can assume that the sentences are about now, since they're in the present tense. So each is true or false. That is, we can treat each as a proposition. So consider:

(6) John is a man and Mary is a woman.

Our use of "and" is fairly regular in English when it is used in this way, coming between two sentences. Usually, though not always, we take it to mean that both sentences are true, so that (6) is true iff both "John is a man" is true and "Mary is a woman" is true. In that case, (6) is a proposition that is formed from two other propositions by joining them with "and."

We can also form:

Mary is not a woman.

John doesn't love Mary.

It's not the case that John is a man.

Our use of "not" or "no" is irregular, sometimes attached to a verb, sometimes appearing as an auxiliary, sometimes prefacing the sentence. But generally it is used to transform one proposition into another that has the opposite truth-value. Thus, "Mary is not a woman" is true iff it's not the case that "Mary is a woman" is true.

There is no scheme vs. proposition ambiguity in our use of these connectives in English. Each creates a new proposition from one or more other propositions.

We create a formal logic when we abstract from these ordinary connectives, writing $\wedge$ for "and" and $\neg$ for "not". We regularize the use of the latter by taking it to precede the sentence it is transforming, as in "Not: John is a man." We then take what are called *propositional variables* to stand for any sentences we might want to reason about as propositions. So we have $p_0, p_1, p_2, \ldots$ going on as far as we wish. We make clear the rules for how to form a new proposition from old ones using connectives, so that we have in the *formal language*, for example:

(7) $p_0 \wedge p_1$

$\neg p_{36}$

$p_{189} \wedge \neg (p_{23})$

$\neg (p_{4318} \wedge \neg (p_2))$

$\neg\neg\neg (p_{4318} \wedge \neg (p_2))$

$(p_{4318} \wedge \neg (p_2)) \wedge p_{800}$

Such formal inscriptions are not true or false. They are schemes in the sense of giving us the forms of the propositions that we can investigate with this formal logic. They are *formal schemes of propositions*.

To have propositions we can reason about with this logic we must *realize* the propositional variables as propositions. Thus, we might take p_0 to be realized, that is stand for, "John is a man" and take p_1 to be realized, that is stand for, "Mary is a woman." These are the *atomic* propositions we are considering. Then $p_0 \wedge p_1$ is realized as "John is a man $\wedge$ Mary is a woman," which is a formal version of (6). If we realize some or all of the propositional variables, that is, if we have a list of which propositions are assigned to which propositional variables, then we can take the formal schemes with those assignments to be the *semi-formal language*. Corresponding to the forms at (7) we might have in the semi-formal language:

(8) John is a man $\wedge$ Mary is a woman

 $\neg$ Ralph is a dog

 Sheila is a herring $\wedge$ $\neg$ (Romulo is a wolf)

 $\neg$ (Edgar is a dog $\wedge$ $\neg$ (John loves Edgar))

 $\neg\neg\neg$ (Edgar is a dog $\wedge$ $\neg$ (John loves Edgar))

 (No one who hates dogs is a good person $\wedge$ $\neg$ (Mary is a dog))
  $\wedge$ John loves Mary

Still, these are not propositions until we say how we will interpret the formal symbols. In classical propositional logic, we say that if A and B are semi-formal propositions, then A $\wedge$ B is true iff both A and B are true; $\neg$A is true iff A is not true. Then each of the semi-formal sentences at (8) is a proposition, either true or false. We can actually determine which they are if we know the truth-values of the atomic propositions in our realization. If we list out those truth-values, we have a *model*. This analysis gives us a way to reason with our atomic propositions using the connectives "and" and "not" in a clearer, more rigorous way than in ordinary English. There is no confusion of scheme and proposition.

In our reasoning we might not know whether a particular atomic proposition such as "John loves Mary" is true or false. Or we might be interested in evaluating whether one proposition follows from one or more others. Letting $A_0, A_1, \ldots, B$ stand for any propositions, we say:

$A_0, A_1, \ldots$, therefore B is a *valid inference* if there is no possible way the world could be such that all of the *premises* $A_0, A_1, \ldots$ are true and the *conclusion* B is false.

We can investigate whether an ordinary language inference is valid by considering its formalization. Thus, we might ask whether the following is valid:

(9) Ralph is a dog.

 $\neg$ (Ralph is a dog $\wedge$ Ralph is a cat)

 Therefore, $\neg$ (Ralph is a cat)

Is there any way in which the premises could be true and conclusion false? What do we mean by a "way the world could be"?

In this context, concerned with only the truth or falsity of atomic propositions and propositions formed from the atomic propositions with ∧ and ⌐, a way the world could be is a model, completely determined by which of the atomic propositions are true and which are false. So to say that the inference at (9) is valid is to say that, within the limited context we are considering, there is no model, no way to assign truth-values to the atomic propositions in it in which all the premises are true and the conclusion false.

When analyzing whether an inference such as (9) is valid, we do not take the sentences that realize the propositional variables to be propositions. Either "Ralph is a dog" is true or it is false if it is a proposition. Yet in evaluating (9) we consider ways in which it could be true or could be false. Each of the models we survey in making that determination does take "Ralph is a dog" to be a proposition. But in surveying those models, we take "Ralph is a dog" to be a scheme, await-ing an assignment of a truth-value in order to be viewed as a proposition. The ambiguity of scheme vs. proposition here arises only when we stand back from any particular use of the sentences realizing the propositional variables and look at all possible uses of them in the context we are considering. We have no choice but to do that in order to make an evaluation of whether an inference is valid.

The ambiguity does not arise just because when we assert "Ralph is a dog" we mean it to be a proposition, yet in (9) we are not asserting it. When in our ordinary language we say "Ralph is a dog, therefore Ralph is not a cat," we are using "Ralph as a dog" just as much as when we simply say "Ralph is a dog." We mean it to be a proposition. It is our interest in deciding whether we can conclude that Ralph is not a cat from "Ralph is a dog" that makes us treat "Ralph is a dog" as a sentence that could be true in some circumstances and false in others in evaluating that inference.

We can take "Ralph is a cat" to be a proposition and not know whether it is true or false. We use inferences such as (9) to help us determine which it is. In doing so we do not forget that "Ralph is a cat" is a proposition, but we consider that sentence to be sufficiently mean-ingful to be considered true or false in other ways the world might be.

There are other motives for us to survey all models. For example, we might wish to know whether a sentence such as "⌐((Edgar is a dog ∧ ⌐(Edgar is a dog))" is true or false due solely to its form. When we do so, we are treating as a scheme what we took to be a proposition.

The scheme vs. proposition ambiguity that arises is not introduced by any of our methods of formal analysis but arises from our reflecting on our reasoning. It is clear when we are treating a sentence as a proposition and when we are treating it as a scheme. We treat a sentence that is a proposition as a scheme only when we are concerned with ways in which it could be true.

Classical predicate logic

When we formalize reasoning based on the view that the world is made up of things, and we quantify over things, then it seems there is an ambiguity of scheme vs. proposition. Consider:

> Someone loves Mary.

This is a proposition. We can make clearer our assumption that we're reasoning about things with this sentence by rewriting it as:

(10)   There is something, and that thing loves Mary.

The whole is a proposition. But what is the status of "that thing loves Mary"? It can't be a proposition. It is like "John loves Mary," before references are supplied for "John" and "Mary." When a reference is supplied for "that thing," it becomes a proposition. Over the last hundred and fifty years, logicians have managed to clarify the role of "that thing" in sentences like (10) using formal methods.

First, a formal language is offered in which instead of "that thing" we use *individual variables*: $x_0, x_1, x_2, x_3, \ldots$. Then we focus on only two ways of talking in English about how many things: "some" and "all." We understand "some" to mean "at least one, but possibly more." And we understand "all" as "all, even if there isn't even one." We use the symbol $\exists$ for this reading of "some," and we use the symbol $\forall$ for this reading of "all." Readings of "some" as "at least one and not all" and of "all" as "all and there is at least one" can be devised from these and the other resources of the formal and semi-formal languages.

So we can formalize (10) as:

(11) $\exists x_1 (x_1$ loves Mary$)$

And we can formalize "Everyone loves Mary" as "$\forall x_1 (x_1$ loves Mary$)$." These are propositions since we previously established the reference of Mary.

Consider, then:

(12) $\forall x_1 \exists x_2 (x_2 \text{ loves } x_1)$

This formalizes "Everyone is loved by someone." In this example "x_2 loves x_1" is a scheme, awaiting either references or stipulations that we're talking about some or all in order to be a proposition. But also "$\exists x_2 (x_2 \text{ loves } x_1)$" is a scheme. It is neither true nor false unless we give a reference for "x_1" or stipulate that we're talking about some or all. It is only (12) that is a proposition.

There is no ambiguity. The operators "$\forall x_1$" and "$\exists x_2$" are used only on *open formulas*, that is, formulas where a variable awaits reference or a quantification; those uses of variables we call *free*. A quantifier converts an open formula into another open formula if there is a free variable remaining, and into a proposition if there is no free variable remaining.[2]

Consider, though:

(13) $\forall x_1 \exists x_2 ((x_2 \text{ loves } x_1) \wedge \text{ John loves Mary})$

Here a scheme (open formula) "$(x_2 \text{ loves } x_1)$" is connected by the propositional connective $\wedge$ to a proposition "John loves Mary." Doing so creates another scheme "$((x_2 \text{ loves } x_1) \wedge \text{ John loves Mary})$." But there is no ambiguity. What is a proposition remains a proposition; what is a scheme continues to be a scheme; and the connection of the two is a scheme. In any case, such formulas are always equivalent to ones in which the quantifiers are attached only to open formulas. In this case, the formula is equivalent to:

$(\forall x_1 \exists x_2 (x_2 \text{ loves } x_1)) \wedge \text{ John loves Mary}$

Complicating the syntax a bit, we could eliminate problematic formulas such as (13).

For generality we devise a fully formal language of predicate logic by taking, in addition to the propositional connectives, the variables, and the quantifiers, *name symbols* $c_0, c_1, c_2, \ldots$ and *predicate symbols* $P_0^1, P_1^1, P_2^1, \ldots, P_0^2, P_1^2, P_2^2, \ldots$, where the subscript gives the number of the predicate symbol and the superscript indicates how many variables are needed for it. For example, "— is a dog" is unary and "— loves —" is binary.

When we realize name symbols as names, such as "John" and "Mary," and predicate symbols as predicates, such as "— loves —" and "— is a dog," we have a formula such as (12) of the *semi-formal language*. The *closed formulas*, that is, ones with no free variable, would seem to be the propositions we are investigating.

But there is a significant problem in trying to make clear what we mean by saying that x_1 stands for a particular thing while pointing to it if we are talking about all things. So for a formula such as (12) to be a proposition, we require first that we specify exactly what things we're quantifying over. For example, we could stipulate that we're talking about all animate creatures, or we could stipulate that we're talking about all dogs, or all people, or . . . . It is only when we specify such a *universe* of quantification that (12) can be treated as a proposition.

The formal methods are meant to clarify quantification by making more explicit the truth-conditions of a sentence such as (12). We take a *model* of the semi-formal language to be a universe, specific references for all the names, and a stipulation of which atomic sentences are true. Only now the atomic sentences are not just ones like "John loves Mary" but also ones like "$(x_2$ loves $x_1)$" when references from the universe are supplied for x_1 and x_2. We make explicit our picking out references as when we say "John loves that thing" by allowing any thing in the universe to be a reference for any variable. So if our universe is all animate creatures, then "x_2 loves x_1" is an atomic proposition when x_1 is stipulated to refer to me and x_2 is stipulated to refer to my dog Birta.

Then in the model, we have:

$\forall x_1 \exists x_2 (x_2$ loves $x_1)$ is true

   iff for any reference supplied for x_1, $\exists x_2 (x_2$ loves $x_1)$ is true

   iff for any reference supplied for x_1, and then for some
       reference supplied for x_2, $(x_2$ loves $x_1)$ is true

There is, however, an oddity here with certain formulas when we think of scheme vs. proposition. Consider:

(14) $\forall x_1 \neg \exists x_2 (x_1$ loves $x_2 \wedge \neg(x_1$ loves $x_1))$

Here the propositional connective $\wedge$ is joining not propositions but schemes. Yet in the analysis of the truth-conditions for (14) we do see the propositional connective joining propositions:

$\forall x_1 \neg \exists x_2 (x_1$ loves $x_2 \wedge \neg(x_1$ loves $x_1))$ is true

   iff for any reference supplied for x_1,
       $\neg \exists x_2 (x_1$ loves $x_2 \wedge \neg(x_1$ loves $x_1))$ is true

iff   for any reference supplied for x_1, and then for some reference
supplied for x_2, (x_1 loves $x_2 \wedge \neg (x_1$ loves x_1)) is true

In the last line $\wedge$ joins two propositions because the variables are
supplied with references. We are justified in using the propositional
connectives to join schemes or to join a scheme and a proposition
because in the final line of the semantic analysis a reference must
be supplied for each variable, so that in the end the propositional
connectives do operate on propositions.[3]

One important motive in devising classical predicate logic is to
investigate the validity of inferences. Consider, for example:

(15)   All dogs bark.
       Ralph is a dog.
       Therefore, Ralph barks.

This is valid: there is no way the premises could be true and conclusion
false. That is, it is not possible for "Ralph is a dog" and "All dogs bark"
to be true and "Ralph barks" to be false—at the same time and in the
same way. But in this analysis we are not treating "Ralph is a dog" as a
proposition, for it is false, and there's no possibility about that. Rather,
we are considering ways those same words in that same order understood in the same way might result in a true proposition. That is, in
analyzing whether (15) is valid, we treat "Ralph is a dog" as a scheme,
not a proposition. It becomes a proposition only upon an indication of
a way the world could be, whether that be a particular time in the past,
an imagined time in the future, another "world" at this very time, ...

We can formalize (15), writing A $\supset$ B for $\neg$ (A $\wedge$ $\neg$B): [4]

(16) $\forall x_1$ (x_1 is a dog $\supset$ x_1 barks)
       Ralph is a dog.
       Therefore, Ralph barks.

If we understand a possibility in the context of our formal analyses of
reasoning to be a model of predicate logic, then we can show that (16)
is valid. If "$\forall x_1$ (x_1 is a dog $\supset$ x_1 barks)" is true in the model, then
no matter what object we let x_1 stand for, "x_1 barks" is true. So in
particular, if we let the thing that is Ralph be what x_1 stands for, and
Ralph is a dog, that is "x_1 is a dog" is true, then " x_1 barks" is true,
too. A possibility is a model of the logic. Again, "Ralph is a dog" is
treated as a proposition when we wish to consider whether it is true or
false, and it is treated as a scheme, awaiting a specification of a model,

when we are investigating whether an inference in which it appears is valid.

We have exactly the same situation as with classical propositional logic except that the indications that can be supplied to turn a scheme into a proposition are more ample. What is held constant in making (1) into a proposition is that "John" and "Mary" are names and "loves" is understood the same regardless of the time. What is held constant in (9) is that "Ralph" is a name and "is a dog" is meant to be understood the same.

Yet that is not really accurate, since in many analyses we say that "dog" might mean a creature similar to what we call dogs but which is different in some specified way. One major problem in understanding and using formal logic is deciding what is held constant across different possibilities, that is, across different ways the world could be. But it is a problem not just for formal logic. It shows up already in evaluating whether (15) is valid. We might consider a way the world could be in which "is a dog" means is a walrus, and "barks" means that it has tusks. If we do, then it is really clear that (15) is only a scheme and not a collection of propositions. However, invoking such a possibility in evaluating inferences is never done in ordinary reasoning.[5] We impose some kind of informal limits on what counts as an acceptable interpretation of the words in (15). Making those explicit is rarely done and would seem to be quite difficult. If we allow unlimited scope for what we mean by a possibility, then (16) is no different from:

(17) $\forall x_1 (P_7 (x_1) \supset P_9 (x_1))$

$P_7 (a_1)$

Therefore, $P_9 (a_1)$

This really is a scheme of propositions.

The nature of possibilities is a big subject, which is discussed in "Possibilities and Valid Inferences" in this volume. But whether we treat (16) as a completely general scheme, no different from (17), or as a collection of meaningful sentences, it is clear that the sentences in it are not treated as propositions in analyzing whether (16) is valid. They are treated as schemes, though the limits on how we turn those into propositions might not be clear.

Just as with classical propositional logic, we treat a semi-formal formula as a proposition when we are considering one way the world could be, that is, when we are considering just one model. When we are surveying models in order to investigate the validity of an inference, we treat the sentences of the semi-formal language as schemes. The contexts are different, and there is no confusion of scheme vs. proposition.

Classical modal logic

Formal propositional logics of possibility and necessity have been devised to clarify how to reason with sentences when considering possibilities.

In formal modal logic, there are two operators besides the usual propositional connectives ¬ and ∧ :

◇   meant to be understood as "possibly," "it is possible that," or "it is possibly true that"

□   meant to be understood as "necessarily," "it is necessary that," or "it is necessarily true that"

Consider, then:

(18)   Ralph is a dog.

This is a proposition, a false one, understood about here and now. In the formal syntax we can form:

◇ (Ralph is a dog)

□ (Ralph is a dog)

These are taken to be propositions.

Let's consider the first:

(19)   ◇ (Ralph is a dog)

This is meant to be understood and analyzed as:

It is possible that "Ralph is a dog" is true.

If (18) is a proposition, then it is not incorporated in (19), for then we would have:

It is possible that "Ralph is a dog" is true right here and now.

That's nonsense, for either "Ralph is a dog" is true or it is false now, and there's no possibility about that. In (19) "Ralph is a dog" is a scheme, true in some ways the world could be, false in others.

In the syntax of formal modal logic we can attach the possibility operator or the necessity operator to any sentence that we have in classical propositional logic. Thus, for example, we can form:

◇ (Ralph is a dog ∧ ¬(Ralph barks))

□ (¬(Ralph is a dog ∧ ¬(Ralph barks))

These sentences formalize how we might reason about inferences, considering whether there is a way the premises could be true and conclusion false.

It would seem that there is no confusion of scheme vs. proposition here. Thinking of the system as formalizing how we reason about inferences and possibilities, all the sentences are schemes. So we can iterate the possibility and necessity operators and nest one within another as in:

(20)   □ ◇ (Ralph is a dog)

□ (Ralph is a dog ∧ ◇ (Ralph barks))

All the sentences are schemes, and the possibility and necessity operators transform schemes into schemes. We are never concerned with whether a particular atomic proposition such as "Ralph is a dog" or a particular complex proposition such (19) or those at (20) are true.

But that's just wrong. We are concerned with whether sentences such as those are true. The inference (15) is valid iff the following is false:

(21)   ◇ (All dogs bark ∧ (Ralph is a dog) ∧ ¬(Ralph barks))

Perhaps the formal semantics for such sentences can clarify whether we are talking of schemes or propositions.

Let's consider one particular semantics, the semantics of what is called *classical logical necessity*.[6] We proceed just as we did when using classical propositional logic to analyze whether (15) is valid. We take a possibility to be a model of classical propositional logic, that is, an assignment of truth values to the atomic schemes where ∧ and ¬ are interpreted as in classical propositional logic. We designate one of those models as being the world as it is here and now, the *actual world*. Then we use all of those "sub-models" together to make a *model* by using the following truth-conditions:

□ A   is true   iff   in every possibility, that is in every sub-model,
                            A is true

◇ A   is true   iff   there is some possibility, that is there is some
                            sub-model, in which A is true

Then we evaluate every sentence relative to the actual world. So in this model, "Ralph is a dog" is false. But "◇ (Ralph is a dog)" is true because there is a way we can assign truth-values to the atomic schemes in which "Ralph is a dog" is assigned to be true.

And now we do have a confusion of scheme and proposition. We treat "Ralph is a dog" as a proposition in this model, but we treat that sentence as a scheme in "◇(Ralph is a dog)." And though we treat the latter as a proposition, we treat it as a scheme in:

(22)   □ (◇ (Ralph is a dog))

The truth-conditions for this are:

(23)   □ (◇ (Ralph is a dog)) is true
    iff in in every sub-model "◇ (Ralph is a dog)" is true
    iff in every sub-model, there is some model relative to that
       in which "Ralph is a dog" is true

But there's no "relative to" here; we're always looking at all possibilities. So (22) is equivalent to "◇ (Ralph is a dog)." Any iteration of the possibility and necessity operators can be replaced by the use of a single operator, and sentences with nesting of operators as in the last sentence at (20) are equivalent to ones with no operator within the scope of another. All the formal mechanism reduces to just an investigation of sentences of the form □A and ◇B where A and B are sentences of classical propositional logic. This is a lot of work to get us back to where we were when we first began investigating (15). Only we have not clarified our informal analyses of (15). By trying to formalize that process we have brought together what we had previously kept in separate contexts: the use of a sentence as a proposition in a particular model and as a scheme when surveying models. Formal modal logic does not clarify; it introduces confusion by trying to meld many models into one.

In other modal logics the confusion is worse.[7] A sentence such as (22) cannot be reduced to one that is simpler. The formal semantics give truth-conditions for it in terms of an *accessibility relation* between possibilities, that is between sub-models. So in the truth-conditions at (23) "relative to" does matter. We are talking about schemes being true in our actual world by considering how they are true in other models, which requires considering other models relative to those, and so on. Once again everything appears to be a scheme, except we really do want to know whether (21) is true.

Worse, we have no way to say whether the formal semantics are apt because we have no prior intuition, indeed no idea at all of the conditions under which (22) is true. We simply don't talk that way. In our ordinary speech we would consider "It's necessary that it's possible that Ralph is a dog" to be nonsense. Yes, we can put the words together in that way, but they make no sense. The formal modal logic is pointed to as a way to clarify what such a sentence means, allowing philosophers to feel confident in their analyses of possibilities. But that confidence is built on a serious confusion, an attempt to meld into one system the use of sentences as propositions and the investigation of whether inferences are valid.

Often it's said that formal modal logic is based on a use-mention confusion. In "It's possible that Ralph is a dog" the sentence "Ralph is a dog" is mentioned not used, and what we should really say is "It's possible that 'Ralph is a dog' is true." But that's wrong because in "It's possible that Ralph is a dog" the proposition "Ralph is a dog" is neither used nor mentioned. The problem, which appears to be at the heart of formal modal logic, is a confusion of scheme vs. proposition.

Arthur Prior developed his approach to formal temporal logics in conscious analogy with formal modal logics. So let's return to the problem of scheme vs. proposition in his work.

Resolving the ambiguity in Prior's approach to temporal logic?

Recall that in Prior's approach to temporal logic we take "John loves Mary" as a proposition, yet in the proposition "P (John loves Mary)" we take that sentence to be a scheme. And the latter is a scheme not a proposition in "FP (John loves Mary)." The approach does not segregate uses of sentences as propositions and uses of sentences as schemes. Can we resolve this ambiguity?

We do not form a new proposition "FP (John loves Mary)" from the proposition "P (John loves Mary)." Rather, what we have, and what the formal semantics assume, is that each of P, F, PP, PF, FP, FF, PFP, ... is a distinct "operator" that when prefixed to a proposition-scheme forms a proposition. Thus, each of "John loves Mary," "P (John loves Mary)," "F (John loves Mary)," "PF (John loves Mary)," and so on is an atomic proposition. In the usual formulations of a propositional logic, an atomic proposition is taken to be a unitary whole with no internal structure. In this approach an atomic proposition does have internal structure: a proposition-scheme + tense marker. The axioms of a temporal logic in Prior's tradition can then be understood as relating propositions in terms of the internal structure of not just compound but also atomic propositions.

Before we go further, we should ask why we would want to have such a proliferation of temporal scheme-into-proposition operators.

One view of logic is as a guide to reasoning well. We investigate how we talk, how we reason, and we resolve ambiguities, clarify inferences, and sometimes see more deeply into how our assumptions about the world affect our reasoning. Formal tools are seen as abstractions from our ordinary reasoning, in some cases helping us to find what we have good reason to believe. If that is how we view logic, how do we arrive at Prior's approach? Neither in English nor in any Indo-European language do we say "In the past John loves Mary." Prior says it is only an accident of our grammar that we don't treat "in the past" and "in the future" as adverbs of sentences.

> I want to suggest that putting a verb into a past or future tense is exactly the same sort of thing as adding an adverb to the sentence. "I *was* having my breakfast" is related to "I am having my breakfast" in exactly the same way as "I am *allegedly* having my breakfast" is related to it, and it is only an historical accident that we generally form the past tense by modifying the present tense, e.g. by changing "am" to "was", rather than by tacking on an adverb. In a rationalized language with uniform constructions for similar functions we could form the past tense by prefixing to a given sentence the phrase "It was the case that", or "It has been the case that" (depending on what sort of past we meant), and the future tense by prefixing "It will be the case that". For example, instead of "I will be eating my breakfast" we could say
>
> > It will be the case that I am eating my breakfast,
>
> and instead of saying "I was eating my breakfast" we could say
>
> > It was the case that I am eating my breakfast.
>
> The nearest we get to the latter in ordinary English is "It was the case that I was eating my breakfast", but this is one of the anomalies like emphatic double negation. The construction I am sketching embodies the truth behind Augustine's suggestion of the "secret place" where past and future times "are", and his insistence that wherever they are, they are not there as past or future but as present. The past is not the present but it *is* the past present, and the future is not the present but *is* the future present.

There is also, of course, the past future and the future past. For these adverbial phrases, like other adverbial phrases, can be applied repeatedly—the sentences to which they are attached do not have to be simple ones; it is enough that they be sentences, and they can be sentences which already have tense-adverbs, as we might call them, within them. Hence we can have such a construction as

> It will be the case that (it has been the case that
> (I am taking off my coat)),

or in plain English, "I will have taken off my coat". We can similarly apply repeatedly such *specific* tense-adverbs as "It was the case forty-eight years ago that".[8]

Prior is trying, it seems, to clarify how to reason about time by treating "in the past" and "in the future" analogously to how we use "it is possible that" and "it is necessary that." Yet his approach is hardly a clarification if it introduces more confusion than we had before.

Still, if I understand correctly, there are some languages that treat "in the past" and "in the future" as operators on sentences. In Chinese, in particular, a phrase roughly translated as "before" can precede a sentence, a paragraph, or even a whole story. In American Sign Language words like "yesterday" or "next month" can precede a sentence, as in "Yesterday, John loves Mary," though the more general operators of "in the past" and "in the future," again if I understand correctly, attach only to verbs. But I know of no language in which an operator like "in the past" or "in the future" can preface a sentence that already begins with one of those. Iterations are not allowed.

And why would we want to use such iterations? We do so in order to move around the time relative to which the operators are meant to pick out a time, as in (4) explicated in (5). But we never change the relative time all by itself in ordinary speech. We do it only for one sentence relative to another. Consider:

(24)   Mary had loved Hubert before John loved Mary.

The truth conditions for this are:

> There is a time in the past at which "John loves Mary" is true, and, relative to that time, there is some time in the past at which "Mary loves Hubert" is true.

This we can model in Prior's approach, though not by iterating temporal operators. Rather, we would use:

(25)   P (John loves Mary ∧ P (Mary loves Hubert))

The conditions for this to be true are:

> At some time in the past "John loves Mary" is true and, relative to that time, there is some time in the past at which "Mary loves Hubert" is true.

In the formal models, we have:

> P (John loves Mary ∧ P (Mary loves Hubert))   is true
>
> iff   there is a time $t < n$ such
> $M_t \models$ John loves Mary ∧ P (Mary loves Hubert)
>
> iff   there is a time $t < n$ such
> $M_t \models$ John loves Mary and $M_t \models$ P (Mary loves Hubert)
>
> iff   there is some time $t < n$ such
> $M_t \models$ John loves Mary and some time $t' < t$ such that
> $M_{t'} \models$ Mary loves Hubert

But here once again we have an ambiguity of scheme vs. proposition. We have agreed to take the temporal operators as converting schemes to propositions, yet in (25) "P (Mary loves Hubert)" has to be understood as a scheme.

If this is a major motive for using Prior's approach to formal systems of temporal logic, we need to ask whether the scheme vs. proposition confusion is really needed or whether there are more straightforward, simpler ways in which we can formalize reasoning with propositions such as (24).

Tenses for predicates

Let's start by considering another approach to temporal logic that I present in *Logic, Language, and the World, Volume 2: Time and Space in Formal Logic*. My goal there is to show how to formalize reasoning from English and similar languages. In English we have a part of speech we call an *infinitive*. We do not use an infinitive as the main verb in a sentence. Rather, we add a tense to the infinitive using a suffix and/or auxiliary words to get what we call a *predicate*. For example, in English from the infinitive "to talk" we have:

talk(s)	simple present
talked	simple past

will talk	simple future
is talking	present progressive
was talking	past progressive
will be talking	future progressive
have (has) talked	present perfect
had talked	past perfect
will have talked	future perfect
have (has) been talking	present perfect progressive
had been talking	past perfect progressive
will have been talking	future perfect progressive

Though there are many irregularities in forming tensed verbs in this way, and many variations, these are the twelve basic ways to form a predicate from an infinitive in English.[9]

In English, "John to talk" is not a sentence, nor (does it represent) a proposition. But "John talked," "John talks," and so on are sentences. They are (or represent) propositions when an indication is made of what counts as the present, or what counts as the present in the past, or what counts as the present in the future.

We can formalize the use of tenses in this manner. In predicate logic we include in the formal language symbols for infinitives, $I_1, I_2, I_3, \ldots$. We also include in our formal language the following *tense markers*:

simple present
simple past
simple future
present progressive
past progressive
future progressive
present perfect
present perfect progressive
past perfect
past perfect progressive
future perfect
future perfect progressive

Then for each *n* the following is a predicate symbol:

I_n-simple present

I_n-simple past

I_n-simple future

I_n-present progressive

I_n-past progressive

I_n-future progressive

I_n-present perfect

I_n-present perfect progressive

I_n-past perfect

I_n-past perfect progressive

I_n-future perfect

I_n-future perfect progressive

To utilize this formalism for reasoning, we realize some or all of the infinitive symbols. For example, we can realize I_1 as "to talk," I_2 as "to run," I_3 as "to give," I_4 as "to take."[10] A complex of such a realization with a tense marker then plays the role that a tensed infinitive does in English and that a predicate does in the realization. Examples of sentences in a semi-formal language are then:

Ralph (to bark-present)

Lemuel (to give-past perfect)

$\forall x_1$ (x_1 (to give-past perfect) $\rightarrow$ $\exists x_2$ (x_2 (to take-past)))

In the usual formulation of predicate logic, an atomic predicate is taken to be a unitary whole with no internal structure. In this system an atomic predicate does have internal structure: an infinitive + tense marker. We can create a formal temporal logic by formulating axioms relating propositions in terms of the internal structure of atomic predicates, as I do in *Logic, Language, and the World, Volume 2: Time and Space in Formal Logic*.

Were we to view logic not as a guide for how to reason well but as a formal system that codifies truths about the world, whether abstract or not, we might choose instead of these twelve basic tense markers some more general approach to tenses. We could have markers:

Present, P, F

Then, for example, we could take as a predicate any of:

to talk-Present

to talk-$P^{i}1F^{j}1P^{i}2F^{j}2 \ldots P^{i}nF^{j}n$

to talk-$F^{i}1P^{j}1F^{i}2P^{j}2 \ldots F^{i}nP^{j}n$

where $i_1 \neq 0$ and

$$P^0 = F^0 = \text{nothing}$$
$$P^1 = P \qquad\qquad F^1 = F$$
$$P^{n+1} = PP^n \qquad F^{n+1} = FF^n$$

More generality could be gained by adding progressive markers. It is clear, I hope, how we could proceed to give a formal language and define realizations of that.

This approach would be similar to the way of resolving the ambiguity in Prior's approach by taking P, F, PF, FP, PP, FF, . . . as distinct scheme-into-proposition operators. But here there is no confusion of scheme and proposition. There is, just as in our ordinary speech, propositions about times relative to other times which can be joined with connectives such as "and" and "not." We could formalize (24) with:

Mary (to love-past perfect) Hubert $\land$
    John (to love-simple past) Mary

That may be right. But it misses a key part of (24): the connective "before."

Temporal propositional connectives

Esperanza Buitrago-Díaz and I in "A Propositional Logic of Temporal Connectives" have shown how to formalize the use of temporal connectives like "before." We start with the usual propositional connectives $\neg$ and $\land$ and sentences that are to be taken as propositions in a model, just as in classical propositional logic. We add to this language connectives meant to formalize (roughly) the ordinary language connectives "before," "after," and "at the same time as." Thus, we might have:

(26)   Mary loves Hubert before John loves Mary.

For a model of the logic, we first take a collection T of instants that is linearly ordered to be the timeline. Each atomic sentence is meant to be about a particular time, which is an interval of T; that is, it is meant to describe the world at that time. For example, we could assign the entire day January 18, 1953 to "John loves Mary"and assign to "Mary loves Hubert" the year 1946. Then for each atomic sentence we say whether it is true or false of the time assigned to it. The connectives ⌐ and ∧ are evaluated classically, and we take, roughly:

p before q is true
   iff the time assigned to p is before the time assigned to q and both p and q are true

p after q is true
   iff the time assigned to p is after the time assigned to q and both p and q are true

p at the same time as q is true
   iff the time assigned to p is the same as the time assigned to q and both p and q are true

For example,

"Mary loves Hubert before John loves Mary" is true

iff the time assigned to "Mary loves Hubert" comes before the time assigned to "John loves Mary," and both "Mary loves Hubert" and "John loves Mary" are true of those times

There is only one model. There is no ambiguity of scheme vs. proposition. There are only propositions and propositional connectives. A proposition is about some time, and it is either true or false.

But (26) does not formalize (24), for (26) does not require that the times assigned to the constituent propositions be in the past in order for the whole to be true. To accomplish that, we need to designate one or a collection of atomic propositions as being about the present. Perhaps "Spot is barking" will do. Then noting that "⌐ (Spot is barking ∧ ⌐ Spot is barking)" is true no matter the time assigned to "Spot is barking," we can say that a proposition p is about the past if the following is true:

⌐ (p ∧ ⌐p) before ⌐ (Spot is barking ∧ ⌐ Spot is barking)

We can abbreviate this as P(p).

Similarly we can have abbreviations F(p) and Present (p) that are true, respectively, if p is about a time in the past, or is true about a time in the future, or is true about the present. These are abbreviations, ways for us to say that a proposition is about the past. For example, consider:

John loves Mary ∧ P (John loves Mary)

This is true iff John loves Mary at some time in the past. Superficially there is a similarity to what is done in Prior's approach, but there is no ambiguity, no confusion. We are making propositions from other propositions, and we can stand outside the system to note whether those are about the present, the past, or the future. After all, what is present, past, or future in a language of only relative time markers must be designated from outside the system, which we do here by designating a particular proposition or collection of propositions as about the present.

If we add these temporal propositional connectives to the predicate logic of tensed atomic predicates as described above, we can then formalize (24):

Mary (to love-past perfect) Hubert before
    John (to love-simple past) Mary

The syntax is clear and unambiguous. The formal semantics are clear and unambiguous. And we can add axioms governing the relations of sentences with tensed atomic predicates to temporal connectives. We can even prove completeness theorems for the logic of tensed atomic predicates, for the logic of temporal connectives, and for the two logics melded together.

We can clarify and give a guide to how to reason well. Or, if you consider logic to be a description of the most general truths of the universe, we can see our way to an "intellectual intuition" of those truths that are formalized in the logics. There is no impediment of confusion to stop us.

Quantifying over times

Another way to formalize reasoning that takes account of time is to use specific time markers such as January 6, 1947, or 2:55 p.m. March 3, 2013, or 306 BCE. Let's assume that we have a timeline composed of instants and names for some of those.

To what do we attach such markers? Consider the sentence:

John loved Mary on June 6, 1970.

We could formalize this by attaching a time-marker to the sentence as a whole as in:

(John loves Mary) (June 6, 1970)

But to do so leads us into the same confusion of scheme vs. proposition we had with Prior's approach to temporal logic.

Instead, as in *Time and Space in Formal Logic*, we can attach time markers to predicates. That is, a unary predicate such as "— to talk" is now construed as a binary predicate, "— to talk (—)," where the blank in parentheses is to be filled with a time marker. Similarly, "— loves —" is now construed as a predicate with three blanks to be filled "— loves (—) — ," the one in parentheses to be filled by a time marker. So we could have the semi-formal sentence:

John loves(June 6, 1970) Mary

We don't need to use tenses or any relative time markers in this system because we can compare the time markers. Thus, "June 6, 1970 is before June 6, 1982" would be one of the sentences of the semi-formal language. But not all the times we want to talk about have names. So we introduce variables that are meant to range over times: $t_0, t_1, t_2, t_3, \ldots$. Then a semi-formal proposition might be:

$$\exists t_1 \, \exists x_3 \, (\text{John loves}(t_1) \, x_3)$$

To formalize talk of the past and the future we have to designate a time marker for the present, say "March 3, 2013." Let's call that n. Then we can formalize

John loved Mary

as

$$\exists t_1 \, (\, (\text{John loves}(t_1) \, \text{Mary}) \wedge t_1 \text{ is before } n)$$

We don't speak like this in ordinary English. But the method is close to how we sometimes incorporate time indications in our speech, and it is clear and easy to deal with formally.

To give a formal logic along these lines we have to adopt axioms for how times relate, whether time is linear, for example, or branching. We have to adopt axioms for how the time markers and variables interact with the names and variables for objects. Such a logic for quantifying over times is clear, creates no scheme vs. proposition confusion, is close enough to ordinary speech for us to have clear intuitions about how to proceed in giving truth-conditions for sentences, and helps us uncover assumptions about how we understand time and objects in our reasoning.

Conclusion

Prior devised his approach to tense logic in conscious analogy with formal modal logics.

By trying to coalesce many models of classical propositional logic into one model, modal logics create a scheme vs. proposition confusion. It is not clear what is true, what is true relative to, nor what is the status of any sentence in the semi-formal language. The many analyses of possibility and necessity that are made using formal modal logic, analyses that purport to clarify our assumptions about the nature of what is possible and what is necessary, are all suspect, based as they are on an underlying confusion of proposition vs. scheme that infects all the work.

That confusion is much worse in the formal systems of temporal logic built on Prior's work. There is no need to coalesce many models of classical propositional logic into one when taking account of time in our reasoning. To do so leaves us with no clear idea whether we are talking of scheme or proposition. The many analyses of problems in reasoning that first Prior and then others made using his systems of temporal logic are suspect, based as they are on an underlying confusion of proposition vs. scheme that infects all the work.

From the very beginning of Prior's work on temporal logic this confusion of scheme and proposition is evident. In "Tense-Logic and an Analogue of S4" in *Time and Modality*, Prior begins by taking sentences such as (1) to be schemes of propositions:

> In the logic of tenses, the ordinary statement-variables *p, q, r,* etc., are used to stand for statements in what is not now the ordinary sense of the term "statement", though it was the ordinary sense in ancient and medieval logic. They are used to stand for "statements" in the sense in which the truth-value of a statement may be different at different times .... The statement "It will be the case that Professor Carnap is flying to the moon", as I understand it, is not a statement *about* the statement "Professor Carnap is flying to the moon", but a new statement about Professor Carnap, formed from the simpler one by means of the operator *F*. p. 8

But then Prior begins to talk of those sentences as propositions:

> The idea that it is *necessary* to introduce a special present-tense operator would, moreover, have extremely awkward formal consequences. For to say that such an operator is necessary is to say that the expressions to which we attach it would not be propositions,

at all events not tensed propositions, without it. This in turn is to say that tense operators do not form propositions out of propositions, at all events out of tensed propositions; rather, they form propositions out of merely juxtaposed nouns and verbs, or they form tensed propositions out of untensed ones. And from this in turn it would follow that tense operators cannot be iterated or attached to propositions to which tense-operators are already attached; that is, we would have to rule out such forms as "It will be the case that it has been the case that *p*". And to rule this would be practically to destroy tense logic before we have started to build it. p. 10

I agree. But I take this to mean that tense logic as he does it is wrong, not that we must persevere in the mistake.

We talk lots of nonsense and confusion. We often reason badly. As logicians we try to bring clarity to our reasoning. We have a responsibility not to add to our confusion.

Dedicated to Esperanza Buitrago-Díaz

Notes

1. (p. 99) The discussion that follows can be modified to apply to propositions taken to be abstract objects. The issue then is whether the sentence under discussion stands for, represents, points to, or somehow indicates a proposition. Similar comments apply to all the formalisms discussed here.

    The discussion can also be modified to apply to reasoning with many "truth-values" by using the dichotomy of designated vs. undesignated as a true/false division, as I discuss in "Truth and Reasoning" in this volume.

2. (p. 107) Most logicians allow in their formalisms sentences such as

$$\forall x_2 \, \exists x_1 \, (x_1 \text{ loves } x_1)$$

where $\forall x_2$ prefaces a proposition. But that is only for convenience in presenting the syntax. Such superfluous quantifications are not needed, as I show in *Logic, Language and the World, Volume 1: The Internal Structure of Predicates and Names*. Without superfluous quantification it is always clear whether we are dealing with a scheme or a proposition (relative to a model), and at no point in the work do we require a sentence to be read as a scheme in one context and a proposition in another.

    Some logicians allow "$(x_1 \text{ loves } x_2)$" to be read as a proposition, under-standing it to mean "for all x_1 and all x_2 (x_2 loves x_1)." But to do so does create a confusing ambiguity of scheme vs. proposition.

3. (p. 109) See *Predicate Logic* for a fuller discussion.

4. (p. 109) Why this formalization is apt is explained in *Predicate Logic*.

5. (p. 110) See my *Critical Thinking*.

6. (p. 112) That is, **S5**.

7. (p. 113) See my *Propositional Logics* for a full development of those and the logic of classical logical necessity, **S5**.

8. (p. 116) "Changes in events and changes in things," pp. 40–41.

9. (p. 118) See *Understanding and Using English Grammar* by Betty Schrampfer Azar.

11. (p. 119) I discuss only unary predicates here. *Logic, Language, and the World*, Volume 2: *Time and Space in Formal Logic* deals with predicates of any arity.

The Timelessness
of Classical Predicate Logic

Classical predicate logic takes no account of time. It was created to formalize reasoning in mathematics in which mathematical claims are viewed as timeless: not eternally true, not true in the past, present, and future, but timeless.

In classical predicate logic no account is taken of the times at which the things in the universe of a model are meant to exist. They all exist in a timeless status; the coming into existence and going out of existence of them is of no concern. The only existence we reason about in classical predicate logic is that which is suitable to allow for a thing to be the value given to a variable by an assignment of references.

Reference in classical predicate logic is timeless. We say that we can assign Socrates to x, though Socrates does not exist now. This is not the issue of whether we have the ability to pick out a thing, for we think that we can pick out Socrates among all other things to be the reference of a variable.[1]

Quantification is timeless, too. For an existential quantification to be true, there must be something in the universe that can be assigned to the variable that makes the resulting proposition true, and that assignment is timeless. For a universal quantification to be true, each thing in the universe, regardless of any considerations of time, must satisfy the predicate.

Because of this timeless nature of the semantics of classical predicate logic, we cannot use it to formalize reasoning that takes account of time. For example, consider the following archetype of a valid inference:

(1) All dogs bark.
    Birta is a dog.
    Therefore Birta barks.

The standard formalization in classical predicate logic is:

(2) $\forall x$ ((— is a dog) (x) → (— barks) (x))
    (— is a dog) (Birta)
    Therefore, (— barks) (Birta)

This formal inference is valid, too. We justify that by saying that if the collection of all things that are dogs is within the collection of all things that bark, then if Birta is a dog, she barks. It is remarkably unclear what we mean by this. The reading of the predicates isn't atemporal, since Birta is a thing in time. Nor is the reading of the predicates omnitemporal, since we don't mean that Birta is barking all the time. It's more like ascribing essential attributes or permanent capabilities or dispositions to things. If Birta is a dog, that is, if she has that attribute without any reference to time, then Birta barks, without any reference to time. But that Birta barks is not an essential attribute of Birta. Or perhaps we need to think that it is in order to use classical predicate logic to formalize (1). And perhaps that is indeed what we mean by (1). This is how we must understand the wffs at (2): they are true because being a dog or barking is an attribute we ascribe to an object independent of time.

Suppose we wish to use both "— is a puppy" and "— is a dog" in a semi-formal language. Those predicates are related in meaning: a puppy is an immature dog. So we should adopt a meaning axiom to codify that:

(3) $\forall x$ ((— is a puppy) (x) $\leftrightarrow$
        ((— is a dog) (x) $\wedge \neg$ (— is mature) (x))

What does this mean? It's said to be atemporal, but nothing is a puppy atemporally. The truth-conditions for (3) are: something is in the collection of things that are puppies if and only if it is in the collection of things that are dogs but are not mature. Suppose, then, that my dog Birta is in the universe of a model. Is she in the collection of mature dogs or is she in the collection of puppies? To use (3), we must choose, but that means we can formalize no true proposition about Birta when she is the other.

Whatever properties we ascribe to an object in a model must be unchanging. That's just to say that the atomic predications true in a model are fixed. That does not preclude, however, having all of the following propositions true in a model:

(— was a puppy) (Birta)

(— is a dog) (Birta)

(— is mature) (Birta)

Even with the meaning axiom (3), these are not inconsistent. Such a model amounts to setting out what is true of certain objects at one particular time. In 2009 "(— is a dog) (Birta) ∧ (— is mature) (Birta)" and "(— was a puppy) (Birta)" are both true. But if that's how we interpret our models, then (3) does not ensure that "((— was a dog) (Birta) ∧ ¬ (— was mature) (Birta)" is true in that model. We would have to add as well:

∀*x* ((— was a puppy) (*x*) ↔
        ¬ ((— was a dog) (*x*) ∧ (— was mature) (*x*))

But that is certainly false, for Birta was both a puppy and was a mature dog, just not at the same time. We'd need to find a better way to relate "— was a dog" to "— is a dog," and "— was a woman" to "— is a woman," and . . . . To reason about things in time in classical predicate logic, about how objects have different properties at different times, we have to adopt *ad hoc* meaning axioms governing every atomic predicate.

Classical predicate logic is useful for formalizing reasoning about things outside of time, or about essential attributes or permanent capacities of things that are in time. We need to incorporate temporal aspects of propositions into the semantics of classical predicate logic and ways of talking about that into the syntax in order to reason about things in time. I describe how to do that in the essay "Reflections on Temporal and Modal Logic" in this volume.

Notes

1. (p. 127) This is the way to understand W. V. O. Quine's memorable phrase "To be is to be the value of a variable" in "Designation and Existence," p. 708. Quine didn't mean that snow doesn't exist since it can't be the value of a variable, and he amended that dictum to read on p. 13 of "On What There Is" to read "To be assumed as an entity is, purely and simply, to be reckoned as the value of a variable." In *Predicate Logic* I suggest that predicate logic as a whole characterizes our notion of thing.

Events in the Metaphysics of Predicate Logic

There are significant limitations on what we can formalize from ordinary reasoning in predicate logic. Some have argued that by recognizing that events are things we can overcome some of those limitations. But viewing events as things leads only to confusions in the metaphysical basis of predicate logic.

Predicate logic is based on the assumption that the world is made up of things. We use names to designate particular things; we use variables as temporary names to pick out things; and we use quantifiers to speak of things in general. It is difficult if not impossible to specify exactly what we mean by "an individual thing," even though we agree in many instances. So we start with an intuition, develop our logic, and then, at best, we can say that an individual thing is whatever we can reason about in predicate logic.[1]

We have rules for how to formalize ordinary reasoning into the language of predicate logic.[2] We test our formalizations against our informal intuitions and formal semantics to see whether we can treat dogs, beauty, numbers, and feelings as things. One of the rules for formalizing is fundamental: any ordinary language inference that is valid should be formalized as a valid formal inference, and any invalid one as invalid.

So consider:

(1) Juney is barking loudly.
   Therefore, Juney is barking.

This is valid. Parsing these sentences as we do for predicate logic, we have two unrelated predicates: "is barking loudly" and "is barking." Thus, the schematic form of the inference in predicate logic is:

P(a)
Therefore, Q(a)

Hence, with the assumptions and limitations we have in establishing predicate logic, the formalization of (1) in predicate logic is invalid. With predicate logic we can gain no insight into why (1) is valid.

We can, however, extend predicate logic to allow for formalizing (1) and similar inferences that involve adverbs by considering the internal structure of predicates. In predicate logic, as it is usually formulated, a predicate such as "is barking loudly" is atomic, without structure. We can extend that formalism to allow for modifiers of predicates, reading "is barking loudly" as a complex predicate: "is barking" modified by "loudly," as I do in *Logic, Language, and the World, Volume 1: The Internal Structure of Predicates and Names*.

Others have proposed a different way to allow for formalizing inferences like (1) in predicate logic. No new formalism is needed. We only need to recognize that events, like the burning of a flame in a fireplace, Caesar being stabbed, and Juney barking are things. By "recognizing" that events are things, we can formalize (1) as:

$\exists x$ (x is a barking $\wedge$ x is by Juney $\wedge$ x is loud)
Therefore, $\exists x$ (x is a barking $\wedge$ x is by Juney)

This is valid in predicate logic. For example, Terence Parsons in *Events in the Semantics of English*, p. 6, writes:

> The basic assumption is that a sentence such as
>
> > Caesar died
>
> says something like the following:
>
> > For some event e,
> > e is a dying, *and*
> > the object of e is Caesar, *and*
> > e culminates before now.

Formalizing in this manner there is no need for an analysis of adverbs: event-talk converts adverbs into adjectives. Indeed, there is no need for an analysis of verbs or for recognizing the possibility that processes are not things. Every verb is replaced by a gerund acting as a noun (or in some languages, like Latin, with an infinitive), so the only verb left is the copula of being: "to be" declined in all tenses.

If we formalize along these lines, then propositions we previously viewed as atomic in predicate logic will be parsed as compound, requiring quantification with a variable that ranges over events. A two-sorted logic will be needed to distinguish the use of such variables from variables we use to refer to other kinds of things that would normally comprise the universe of a realization.[3] Thus, though "Juney barks" would seem to be atomic, we should understand it as having a hidden variable and quantifier: "∃*e* (*e* is a barking and *e* is by Juney)." Advocates of this view have to argue that this is how we should understand such propositions, for there is no good evidence that we actually do understand them this way.[4]

The justification offered for viewing events as things is that it allows us to formalize a great deal more in predicate logic than we could before. As Parsons says on p. 146 of his book:

> I don't cite these results as evidence for the theory, or even as phil-
> osophically desirable consequences. The evidence for the theory lies
> in its ability to explain a wide range of data better than other existing
> theories. The existence and nature of events and states are by-products,
> in the same way that the symmetry of space and time are by-products of
> investigations in physics.

This is no justification; it is an example of the fallacy of inference to the best explanation.[5] That good explanations come from particular metaphysical assumptions cannot justify those assumptions. In any case, Parson's comparison to physics is spurious: it isn't the symmetry of space and time that should be compared to whether events exist but the existence of space-time. Nor is it obvious that this is the best way to "account for the data" compared to the method of formalizing adverbs as predicate modifiers that requires no additional metaphyical assumptions.

There are many problems with adopting events as basic to our metaphysics.[6] One, however, is so great that there seems to be no way to proceed in treating events as things in predicate logic: we have no way to distinguish events. We have no way to pick out one event rather than another when we wish to give a reference to a variable, yet doing so is essential for the semantics or use of any predicate logic. When we say that "*x*" is to refer to the stabbing of Caesar, is that the same event as the stabbing of Caesar with a knife? Is it the same event as the stabbing of Caesar with a knife by Brutus? When did the event start? With Brutus conceiving of the action? With Brutus lifting his hand? With Brutus pushing the knife into Caesar? If the last, how far into Caesar did the knife go in that event?

The only way anyone has been able to individuate events is by invoking propositions that are meant to describe them. Thus, each of the above is a different event because each of the following is a different proposition: "Caesar was stabbed," "Caesar was stabbed with a knife," "Caesar was stabbed with a knife by Brutus," "Caesar was stabbed with a knife by Brutus only when Brutus lifted his hand with the knife," . . . .[7] No other way has been presented that is clear enough to use as the basis of naming—that is, picking out reference—in our models. Thus, to understand how to reason with events, we need to know already how to parse and hence how to reason with the propositions that were supposed to be explicated by rewriting them by appeal to hidden variables ranging over events.[8]

Donald Davidson enters into a similar circularity. In "Causal Relations" he says that events are needed to clarify and to give the truth conditions of causal claims, since we apparently talk of events in our ordinary speech. He says we need events as things because otherwise we wouldn't be able to give the logical form of causal claims, meaning a predicate-logic form. But after surveying all the possibilities for criteria of individuating events, in "The Individuation of Events" he comes to the conclusion that the best criterion we can muster is that events are different if and only if they differ in their causes and/or effects. That is, we need events to explain cause and effect, but we first need to understand causes and effects to be able to distinguish events. However, if we cast talk of events as talk of propositions we can analyze cause and effect quite well, as I show in "Cause and Effect" in *Cause and Effect, Conditionals, Explanations* in this series of books.

Robin Le Poidevin in "Relatonism and Temporal Topology" also believes that events are important for cause and effect. He says:

> Not only are events, on the face of it, less mysterious entities than instants, they are clearly things with which we can causally interact. p. 152

But instants are no more mysterious than points in geometry: we arrive at them through a process of abstraction. In any application of a theory of time in which instants play a role, we take as instants intervals of time so short that their duration is negligible relative to the rest of what we are paying attention to.[9] Events, on the other hand, are not obviously entities at all. As Benson Mates pointed out to me, "The cat is on the mat" is supposed to be made true by or describe an event. But what is included in that event? The cat is touching the mat? The cat is upon the mat? The earth, because up and down can only be determined relative to that? Where do we stop? Probably only at the entire universe. But simply, the event is that the cat is on the mat. We use "that" to restate the claim.

There is a further problem with taking events as fundamental to the metaphysics of predicate logic that I have not seen discussed before. Consider:

(2) Juney is a dog.

If events are taken as fundamental in predicate logic, then this should be converted into event talk as something like:

There is an event and it is a dogging and it is by Juney.

But that is to convert "is a dog" into a process and then to convert that back to a thing. I've never seen anything like this. Rather, what is assumed is that (2) is already in primitive form. But then we have an assumption that verbs of "activity" and verbs of classification are quite distinct metaphysically, so that every use of a non-copula verb is an implicit quantification while every use of a copula verb isn't. There seems to be no justification for this metaphysical distinction other than following the implicit metaphysics of some particular language, in this case English, which doesn't seem to be based on anything like an understanding of the nature of the world, just accreted habits of speech crystallized in a grammar.

Propositions are meant to describe the world at some place and time, at least if they are not meant as being about abstract objects. To reify the "part of the world" at that place and time and call it "an event" does not clarify, does not expand the scope of predicate logic, but only leads to confusions.

Notes

1. (p. 131) This is how predicate logic is developed in my *Predicate Logic*.

2. (p. 131) See "On Translations" in this volume, pp. 91–94.

3. (p. 132) The use of different kinds of variables is a standard part of the formalism of predicate logic; see, for example, Chapter XIV.H of *Classical Mathematical Logic*.

4. (p. 133) When I asked one advocate of the use of events in formalizing why we should take events as basic, Donald Davidson replied that it's because we talk that way. Yes, we do talk about events sometimes, but not in the way that such a theory requires. We also talk about feelings and unicorns, but that doesn't mean we believe they exist as things.

   Roberto Casati and Achille Varzi explain Davidson's view more carefully and review debates about the nature of events as things in "Events." In Davidson's comments and Casati and Varzi's article, there is no acknowledgement that the talk they are considering is in English. At best, then, they could be said to be discussing the implicit metaphysics of speakers of English.

5. (p. 133) See "Explanations" in *Cause and Effect, Conditionals, Explanations* in this series of books. It also misconstrues the work of physicists, for all that true (enough) consequences of a theory can show us is the range of application of a theory, as explained in "Models and Theories" in *Reasoning in Science and Mathematics* in this series.

6. (p. 133) A general critique can be found in E. J. Borowski, "Adverbials in Action Sentences." I survey problems that arise in taking events as the basis for causal analyses in Appendix 1 of "Reasoning about Cause and Effect" in *Cause and Effect, Conditionals, Explanations* in this series of books.

7. (p. 134) Compare Parsons in *Events in the Semantics of English*:

   > Most events and states are concrete entities, not abstract ones.
   > First, they are located in space. Since Brutus stabbed Caesar in the
   > marketplace, the theory tells us that there was a stabbing, by Brutus,
   > of Caesar, and the stabbing was in the marketplace. pp. 145–146

8. (p. 134) Nicholas Unwin in "The Individuation of Events" presents a survey of this problem. Arthur Prior in *Past, Present and Future*, p. 18, paraphrases talk of events in favor of talk of propositions.

9. (p. 134) See the discussion of points and lines in geometry in my "Mathematics as the Art of Abstraction" in *Reasoning in Science and Mathematics* in this series of books.

Categoricity with
Minimal Metaphysics

Contrary to the views of many logicians, a categorical finite characterization of the natural numbers can be given in which no infinitary assumption nor assumption about the nature of collections is required beyond what are used in first-order logic. This can be accomplished with an extension of first-order logic in which quantification over names is allowed and in which a formalization can be given of "Every object has a name."

Introduction

First-order logic has a significant drawback: we cannot ensure that we are speaking of an intended model. This drawback is especially notable in dealing with the natural numbers where we all "know" what the intended model is.

In second-order logic we can characterize the natural numbers categorically: there is only one model up to isomorphism. But to employ that logic we must make metaphysical assumptions concerning the nature of collections and predicates far beyond those necessary for first-order logic, assumptions that are not universally shared among mathematicians and logicians and that, on the face of it, seem extraneous to the study of the natural numbers.[1]

In the first-order language of arithmetic, $L(S, +, \times, 0)$, we have all the natural numbers represented via their names, $0, S0, SS0, \ldots$. The problem is that we cannot say, "Every object has a name."

It might seem that we could do that with substitutional quantification. But we cannot restrict the universe of a model there to have only named individuals, and validity of wffs as well as semantic consequence are the same for substitutional quantification as for referential quantification.[2]

We can, though, extend the language of first-order logic to allow for quantification over names so that we can formalize "Every object has a name." We can then exhibit a finite collection of wffs in the language of arithmetic whose only model up to isomorphism is the natural numbers. Doing so will refute the commonly held view that an infinitary assumption or an assumption about the nature of collections and predicates is needed in order to accomplish a categorical characterization of the natural numbers.

The logic LN

We start with the usual first-order language of predicate logic, including functions and equality[3]:

$$L(\neg, \rightarrow, \wedge, \vee, \forall, \exists, =; x_0, x_1, \ldots, P_i^n, f_i^n, c_0, c_1, \ldots)$$

where $i, n \geq 1$

We add to the vocabulary:

name variables a_i $i \geq 0$

In the definition of well-formed formulas (wffs), name variables can occur where the usual variables do, and we add the following clauses:

If A is a wff, then $\forall a_i$ (A) is a wff.

If A is a wff, then $\exists a_i$ (A) is a wff.

For any closed term t, let $A(t/a_i)$ be A with every free occurrence of a_i replaced by t.

A model of this *language of quantification over names* is the same as a first-order model of the original language, where '=' is interpreted as identity and only closed wffs are counted as true or false in the model.[4] The definition of satisfaction in such models is the classical first-order definition supplemented with the following clauses:

For any assignment of references to variables, σ:

$\sigma \vDash \forall a_i$ (A) iff $\sigma \vDash A(t/a_i)$ for every closed term t.

$\sigma \vDash \exists a_i$ (A) iff $\sigma \vDash A(t/a_i)$ for some closed term t.

A closed wff is true in a model if and only if it is satisfied by every assignment of references.

We then define the logic of name quantification.

The logic of name quantification   **LN** is the logic of all models of this language. That is, for any collection of closed wffs Σ of the language and closed wff A, $\Sigma \vDash_{\mathbf{LN}} A$ iff A is true in every model in which all the wffs in Σ are true.

For this logic we can prove the strong form of the Downward Löwenheim-Skolem Theorem by the usual construction using Skolem functions.[5]

Theorem 1 If a collection of sentences has an infinite model, then it has a countable model that is a submodel of the given infinite model.

In **LN** we can formalize "Every object has a name" with the following formula:

AN $\forall x_1 \exists a_0 (x_1 = a_0)$

This is true in a model iff every object in the model has a (possibly complex) name in the formal language. We can also formalize "There is some object that does not have a name" with the following formula:

$\exists x_1 \forall a_0 (x_1 \neq a_0)$

These two formalizations illustrate the difference between quantifying over names and quantifying over named individuals, since neither is formalizable in first-order logic using substitutional quantification. Quantifying over names when used with referential quantification is a new tool for formalizing reasoning. It could be used to create a logic that allows non-referring names in which we could formalize "There is some name for which there is no object" as:

$\exists a_1 \neg \exists x_0 (a_1 = x_0)$

And it could be used to give semantics for logics of sense and reference, where a name can have semantic content in addition to reference.[6]

The natural numbers

Let Qn be the formula AN plus the finite axiom system **Q** of Alfred Tarski, Andrzej Mostowski, and Raphael Robinson[7] in the name quantification language $L(S, +, \times, 0)$. Let **QN** be the collection of semantic consequences of Qn in **LN**. Let $\mathbb{N}$ be the natural numbers with successor, addition, multiplication, and zero.

Theorem 2 The finite set of axioms Qn has up to isomorphism just one model, $\mathbb{N}$. So **QN** is the set of wffs of the language of **LN** that are true of the natural numbers.

Since the isomorphism of models can be given via the assignment of names, the proof of Theorem 2 is entirely constructive. No non-constructive assumptions are required for this system. The only place non-constructive assumptions might be necessary is to show that any countable set can serve as the universe of a model for Qn, and that is only if one assumes a non-constructive definition of "countable."

Define **Arithmetic₁** to be the set of wffs of the usual first-order language $L(\,S, +\,, \times\,, 0\,)$ that are true of $\mathbb{N}$. It follows that **Arithmetic₁** = the part of **QN** in which no quantifier over names and no name variable appears. So by Theorem 2, we have the following.[8]

Theorem 3

> **LN** does not satisfy the Upward Löwenheim-Skolem Theorem.
>
> **LN** cannot be recursively axiomatized.
>
> **LN** is not compact.

The first-order scheme of induction is:

Ind_1 $[A(o) \wedge \forall x\,(A(x) \rightarrow A(Sx))] \rightarrow \forall x\,A(x)$
     where A is a first-order wff in which only x is free

Peano Arithmetic₁ is often characterized as the first-order syntactic consequences of **Q** plus Ind_1 in $L(\,S, +\,, \times\,, 0\,)$. In **LN** we gain no new wffs true of the natural numbers in $L(\,S, +\,, \times\,, 0\,)$ by adding Ind_1 to **QN**.

The second-order axiom of induction is:

Ind_2 $\forall X\,[((X(0) \wedge \forall x_2\,(X(x_2) \rightarrow X(Sx_2))) \rightarrow \forall x_1\,X(x_1))]$

To prove Theorem 2 we do not need to assume that Ind_2 is true of the natural numbers. By assuming that the natural numbers with successor, addition, multiplication, and zero is a model of **Q**, we also have a model for Qn.

The single wff *AN* seems as useful in characterizing arithmetic in terms of semantic consequences as the much more powerful and controversial wff Ind_2. The first-order part of **QN** is the same as the first-order part of second-order arithmetic (Ind_2 plus **Q**). What we cannot do in **QN** or in **LN** is set theory. But it is not clear that we want or need a substantial portion of set-theory to formalize arithmetic.

The logic **LN** introduces no metaphysical assumptions that are not already present in first-order logic, since to employ the usual language of first-order logic we are allowed to employ names, and the process of naming is an essential part of the semantics of first-order logic.[9] Names are things, too, and hence objects that can be quantified over. It might seem that quantifying over names violates the restriction that we should not allow talk of the syntax of a formal language within the language itself, which is imposed to avoid self-referential paradoxes.[10] But no contradiction arises since **LN** is consistent for the same reason **Q** is: it has a model. No new entities are demanded, nor are further assumptions about the nature of language and reasoning required in order to employ quantification over names.[11]

These remarks contrast with the long-held view that assumptions about the nature of sets or infinitary assumptions are needed to formalize more of arithmetic than can be done in first-order logic. That view was already well established in 1949 when Julia Robinson said:

> The way in which we use the terms "arithmetic," "arithmetical or elementary theory," and "arithmetical definability" calls for some comments. For example, by the arithmetic (or elementary theory) of integers, we mean that part of the general theory of integers which can be developed without using any notions of a set-theoretical nature; that is, the part of the theory which can be formalized within the lower predicate calculus.[12]

In 1996 Daniel Isaacson continued and refined that view:

> [First-order Peano arithmetic] consists of those truths which can be perceived as true directly from the purely arithmetical content of a categorical conceptual analysis of the notion of natural number. The truths expressible in the (first-order) language of arithmetic which lie beyond that region are such that there is no way by which their truth can be perceived in purely arithmetical terms. Via the phenomenon of coding, they contain essentially hidden higher-order, or infinitary concepts. On this perspective, Peano arithmetic may be seen as complete for finite mathematics. ...

> Those mathematical truths expressible in the language of arithmetic but not provable in **PA** contain 'hidden higher-order concepts', where what is hidden is revealed by the recognition of the phenomenon of coding. What I mean here by 'higher-order' includes the standard

usage for quantification over sets of individuals in distinction to first-order quantification over the individuals themselves. But I also mean to include in the phrase something of the notion infinitary [*sic*], in the sense of presupposing an infinite totality.[13]

First-order logic is considered the paragon for reasoning about mathematics because it can be recursively axiomatized. That virtue is offset by the very considerable flaw that whatever we say about the natural numbers in a recursively axiomatized system must be ambiguous. Because of non-standard models, there is always a different meaning for what we say than what we intended. And one of the first desiderata of good reasoning is that there should be no ambiguity.

In classical first-order mathematical logic we can axiomatize our reasoning, but we cannot characterize recursively, much less finitely, the truths of the natural numbers. In the logic of quantification over names, we can finitely characterize the truths of arithmetic and eliminate ambiguity, but we cannot recursively characterize our reasoning. In classical first-order mathematical logic we are faced with adding more axioms to obtain more truths of the natural numbers. In the logic of quantification over names we are faced with adding more axioms to obtain more valid wffs. To say that classical first-order mathematical logic is nonetheless better because it is more mathematically tractable is only to say that the beauty of the model theory of classical first-order mathematical logic is more important than accurately reflecting or modeling how to reason well.

In second-order logic we can also eliminate ambiguity in talking about the natural numbers, but that is at the cost of introducing controversial and unnecessary metaphysics.[14]

Notes

1. (p. 137) See Chapter X of *Predicate Logic*.

2. (p. 137) See *Predicate Logic*, especially Chapter IV.M.2.
   Substitutional quantification is a confusion. It takes "for all" to mean "for all named objects," which is clearly not what is meant or wanted in first-order logic. It tries to get rid of the use of variables as temporary names, giving them only the role of place-holders. The result is that we could have a model whose universe is all natural numbers in which "$\exists x_1 \neg (x_1 + x_1 = x_1)$" is false if no object other than 0 is named. We get the right truth-value only if we survey all models that have the same universe but differing assignments of reference to the names, which amounts to taking names as variables. That is how validity and semantic consequence are the same as for referential quantification.

3. (p. 138) See *Classical Mathematical Logic*.

4. (p. 138) See *Classical Mathematical Logic*.

5. (p. 139) See Chapter XI of *Classical Mathematical Logic*.

6. (p. 139) For a logic of sense and reference see "Relatedness Predicate Logic" with Stanisław Krajewski, which was first described in Krajewski's "Relatedness Logic." After using these semantics for that logic of sense and reference and developing the material in this paper, I discovered that Saul Kripke in "Is There a Problem about Substitutional Quantification," pp. 334–355, described what appears to be the same semantics, though I have not been able to find work by him or anyone else that investigates this system. Kripke suggested formulating the logic as a 2-sorted one, but to do so would replace quantification over names with quantification over named individuals.
   Hugues Leblanc in "Semantic Derivations" describes several ways to combine referential and substitutional interpretations of quantifiers, though none appear to be equivalent to the system here.
   It was suggested to me that **LN** is equivalent to a logic using infinitely long formulas. But it is not clear what notion of equivalence is meant. The difference in metaphysics makes it seem unlikely that there is a meaning-preserving translation between **LN** and any such logic (see "On Translations" in this volume).

7. (p. 139) See *Computability*, Chapter 21.C or *Classical Mathematical Logic*, Chapter XIV.B.1.

8. (p. 140) Compare Chapter XIV of *Classical Mathematical Logic*. An anonymous referee suggested that by an adaptation of the completeness theorem in Jon Barwise's *Admissible Sets and Structures*, p. 89, it can be shown that the set of valid wffs of **LN** is Π^1_1, and so by Theorem 2 the set of valid wffs must be Π^1_1-complete.

9. (p. 141) See *Predicate Logic*.

10. (p. 141) See the discussion of the Self-Reference Exclusion Principle in *Predicate Logic*.

11. (p. 141) John Corcoran in "Categoricity" presents a system weaker than second-order logic in which a categorical characterization of the natural numbers can be given in terms of semantic consequences of a finite set of wffs. His system adds to first-order logic universal quantification over a single predicate variable in order to formalize induction, allowing for no logical complications of such formulas. Following Richard Montague he calls this a "slightly augmented first-order language". But the semantics for this slight augmentation require substantial metaphysical assumptions about the nature of collections that go beyond those needed for first-order logic, in particular an analysis of what one means by "all subsets" of a given set.

12   (p. 141) "Definability and Decision Problems in Arithmetic," p. 98.

13. (p. 142) "Arithmetical Truth and Hidden Higher-Order Concepts," pp. 202–203, 210.

14. (p. 142) In *Classical Mathematical Logic* I show that we can also categorically characterize the rational numbers in **LN**. And we can "almost" characterize the arithmetic of the real numbers in **LN** with a theory whose only models are real closed fields between the rationals and the reals.

In **LN** we can also investigate other views of the foundations of mathematics. Let **ExpQN** be **QN** in $L(S, +, \times, \exp, 0)$ plus the following axioms for the exponential function:

$$\exp(x, 0) = x$$

$$\exp(x, Sy) = x \times \exp(x, y)$$

It would seem that there is only one model of **ExpQN** up to isomorphism. But that is precisely what David Isles in "Remarks on the Notion of Standard Non-Isomorphic Natural Number Series" denies, pointing out the infinitistic assumptions on which it is based. We can define:

$$x < y \equiv_{\text{Def}} \exists z \, (z \neq 0 \wedge x + z = y)$$

Isles argues that there are models of **ExpQN** consisting of various standard nonisomorphic natural number series where the following formula fails:

$$\forall x \, \forall y \, (x < y \vee y < x)$$

From the finitistic perspective of that paper it would seem that even adding this latter formula would not ensure categoricity. Quantification over names in the formal language should allow a clearer formal analysis of such questions.

Reflections on Gödel's Theorems

A great deal has been written about the significance of Gödel's
theorems on the undecidability of arithmetic and the unprovability of
consistency. Throughout many of those discussions, an unexamined
platonism is assumed which on examination turns out to impede
analysis of the issues involved.

Kurt Gödel proved two major theorems about the foundations
of mathematical logic. One says that given any formal theory of
arithmetic in which there are axioms characterizing at least addition,
successor, multiplication, and which has a name for zero, there is at
least one formula in the language that is true of the natural numbers
but which cannot be proved in the theory, if the theory is consistent.
The second says that the consistency of such a theory cannot be proved
within the theory itself: a more powerful theory is required.[1] Much has
been written about the significance of these theorems, and many of
those views are ably described and discussed in a book by Stanislaw
Krajewski, which he has summarized in a paper.[2] Throughout many of
those discussions an unexamined platonism is assumed.

For comparison, consider the usual argument that predicates cannot
be linguistic:

a. There are only finitely many English words.

b. So there are at most countably many English sentences.

c. Let $Q_1, Q_2, \ldots$ be a list of all predicates that can be derived
   from English sentences that have only a single gap to be filled
   by a name.

d. Each of these predicates either applies or does not apply to
   any other predicate on the list.

e. So the following phrase, call it Q, is a predicate:
   "— is a predicate on our list that does not apply to itself".

f. So Q must be on our list.

g. But then Q applies to itself if and only if Q is a predicate on
   our list that does not apply to itself.

h. This is a contradiction. So Q cannot be on our list.

i. But our list was supposed to contain all predicates.

Therefore, predicates cannot be linguistic.

This is a bad argument. Whatever "countable" means, it must apply to a fixed, static collection of objects or a fixed way of generating objects. That is what is required for step (c). But the collection of all English sentences is not fixed. We coin new words and new names, we use words in new ways that cannot be accounted for by such a list. In particular, the phrase Q could not have been a predicate—satisfying (d)—until we fixed what "our list" was to refer to, which required making the list. So Q is not on our list. But no contradiction ensues from that. Predicates may be linguistic, but they cannot be the open sentences in any fixed, regular, nonexpandable language.

Now compare what Krajewski writes about Gödel's theorems:

> Another standard consequence of the Incompleteness Theorem is expressed by saying that there is no universal, effectively presented, mathematical theory.[3]

I'm not sure what a universal mathematical theory is, but it seems to assume we can talk of all mathematical truths. Any intuitionist, any constructive mathematician, any person who regards mathematics as abstractions from experience will consider that phrase to be meaningless.[4] We know what we mean by all mathematical truths that are in any mathematics book now; that's perhaps a sufficiently precise collection to wonder whether it could be generated by a consistent theory. But all mathematical truths? Ever thought by anyone? And ones still unthought? Ones unthought in what language? As we continue to do mathematics, we coin new words and new names; we use words in new ways that cannot be accounted for by such a collection, says the constructivist or the person who views mathematics as abstraction from experience. It is only platonists who consider there to be such a collection. As in the argument that predicates can't be linguistic, to assume that there is a collection of all mathematical truths to be codified mistakes a platonic reality for a human process.[5]

Krajewski continues:

> Yet a subtlety discovered by Gödel implies that while there is no universal theory that can be known by us, it is not excluded by his discoveries that there could be an effectively presented theory

capturing "subjective" mathematics, that is, the totality of mathematical theorems accessible to the human mind. Still, it seems that there
is no finite description of the natural numbers that we could formulate
and give to a computer to make it behave as if it understood our
notion of number.

To talk of the totality of mathematical theorems accessible to the
human mind again assumes that mathematics is a platonic collection
of theorems waiting to be discovered by us. The second sentence
raises the question of whether we can emulate human behavior with a
machine, and to what extent doing so might constitute understanding.

Krajewski says:

> The above limitations refer to objective mathematics. Gödel did
> not prove that subjective mathematics, that is mathematics that is
> potentially accessible to the human mind, could not be expressible
> as one theory or one algorithm.

Even when considering the subjective aspects of mathematics, a
platonic ideal is assumed: that which is potentially accessible to the
human mind. Many people believe that there is no body of truths that
is mathematics now and forever.

Krajewski considers the idea that it isn't truths we wish to capture
but our abilities to discover them:

> It is not proven that there is no theory or program that is equivalent to
> our mathematical powers.

What is meant by "mathematical powers"? Is that the abilities
Krajewski and I have to prove theorems in mathematics? Or perhaps
the powers of Krajewski, me, and William Thurston? Or of some
larger group of mathematicians? Any psychologist will tell you that
such a collection is far from clearly defined. In any case, Gödel's work
is irrelevant; our mathematical powers lead to false conclusions, so
either there is a machine—an inconsistent theorem-generating machine
—or if it is meant that there is no consistent machine, that's clearly so.

Gödel himself says,

> There may exist (and even be empirically discoverable) a theorem-
> proving machine which in fact is equivalent to mathematical intuition,
> but cannot be proved to be so, nor even be proved to yield only correct
> theorems of finitary number theory.[6]

Certainly if a machine is equivalent to mathematical intuition, no one
could prove that it yields only correct theorems on finitary number

theory since mathematicians don't do that. Mathematical intuition is not something divorced from human beings. Hence, we can only talk about this person's or that person's mathematical intuition. And any machine that is equivalent to such a particular intuition—and it's certainly not clear what "equivalent" means here—would have to make errors and be inconsistent. What Gödel seems to be saying is that there might be a machine that is equivalent to an idealized notion of mathematical intuition. But such an idealized intuition is already in essence a machine. This kind of comment is symptomatic of the view that logic and mathematical intuition are divorced from people who reason.

Concerning the human aspect of mathematics, Krajewski also says,

> We now know that it matters how the assertion of consistency is expressed. This "intensional" aspect is seen by Kadvany as an example of the presence of an historical dimension in mathematics.[7]

We don't need reflections on deep theorems of mathematical logic to note an historical dimension in mathematics. Only people who thought mathematics was a platonic ideal unconnected to people could have imagined there wasn't an historical dimension present throughout mathematics.

One final thought about Gödel's work. Krajewski says:

> [Gödel is] widely seen as "the greatest logician since Aristotle."

Gödel distinguished some conceptions that were confused together: provability and truth. He showed us that the ancient hope of a universal machine to calculate all truths is not attainable. But his few philosophical writings are only metaphors for his platonism, not especially distinguished in the history of analyses of platonism. He did no systematic work on the nature of logic. He never worked on the areas of logic covered in this series of books, such as reasoning about cause and effect, explanations, and generalizing, which form a major part of our understanding of how to reason well. Even as a formal logician, it is not clear his work is superior to that of John Buridan. The evaluation in the quote is a symptom of identifying logic with classical mathematical predicate logic and various offshoots of that subject, denying any conception of logic as the art of reasoning well.

It is one thing to comment on Gödel's platonism. It is quite another to assume platonism as the basic understanding of mathematics and evaluate Gödel's work by that standard only.[8]

Notes

1. (p. 145) See Epstein and Carnielli, *Computability*, for a presentation of the mathematics along with commentary by others.

2. (p. 145) *Twierdzenie Gödla i Jego Interpretacje Filozoficzne: Od Mechanicyzmu do Postmodernizmu* and "Gödel's Theorem and Its Philosophical Interpretations: From Mechanism to Postmodernism."

3. (p. 146) "Gödel's Theorem and Its Philosophical Interpretations: From Mechanism to Postmodernism," p. 105.

4. (p. 146) See the discussion of those views in "Mathematics as the Art of Abstraction" in *Reasoning in Science and Mathematics* in this series.

5. (p. 146) In a personal communication, Krajewski replied, "We need only the weakest possible Platonic assumptions to make those claims. Namely, instead of mathematics we may just talk about Diophantine equations. It is relatively easy to believe that every specific polynomial with integer coefficients either has solutions in the natural numbers or not. It is highly meaningful that Gödel showed that already in relation to such problems (sentences "there is no solution of a (specified) Diophantine problem") so many of the problems connected to the philosophy of mathematics can be discussed. Understanding by machines in this context can be just seen as the existence of an algorithm that solves exactly the same Diophantine problems that we humans can (potentially) solve."

6. (p. 147) As quoted in Hao Wang, *A Logical Journey: From Gödel to Philosophy*, pp. 184–185.

7. (p. 148) Krajewski, p. 105, referring to John Kadvany, "Reflections on the Legacy of Kurt Gödel: Mathematics, Skepticism, Postmodernism," p. 177.

8. (p. 148) A different perspective on Gödel's theorems is given by David Isles. He points out that it is not just the consistency of the axiomatic theory of arithmetic that is necessary to yield that there is a true but unprovable statement of that theory, but also that exponentiation is total, since names in the theory are given in unary notation. The assumption that exponentiation is total is a major abstraction from experience, perhaps even a platonic assumption, so that we can view that assumption rather than the identification of provability with truth as the problem uncovered by Gödel's theorem. See Isles, "Remarks on the Notion of Standard Non-Isomorphic Natural Number Series."

On the Error in Frege's Proof that Names Denote

An examination of Gottlob Frege's proof that names in his system
denote illuminates the nature of induction proofs in formal logic today
as well as the importance of keeping the formation rules for the syntax
distinct from the semantics of the formal language.

In *Die Grundgesetze der Arithmetik*, Gottlob Frege attempted to show
that every name of his Begriffsschrift has a denotation.[1] If he had been
successful, he would have established that his system was consistent, as
each sentence-name would have had a unique denotation: either the
True or the False.

Frege did not think of what he was doing as a consistency proof: the
last thing he says in his Introduction, p. 25 (xxvi), is that it is improbable
his system is wrong. Nonetheless, his proof plays an important role in
Die Grundgesetze der Arithmetik. Culminating in §31, "Our simple
names denote something," it brings together everything done to that
point concerning names, the nature of functions, and courses-of-values.
After his proof in §31 that his names denote, Frege can show in §32
that his sentence-names express thoughts and hence that it's legitimate
to talk of his assertions in the Begriffsschrift as propositions.

We now know that Frege's system is inconsistent, and we can ask
in what way his proof that his names denote is wrong.

I shall concentrate on §31, as it seems easiest to locate a specific
error there. In tracing the genesis of that error, I'll suggest that the
problem is due not so much to the introduction of the notion of a
course-of-values function but to Frege's conception of how function
and object names denote and that there are no simple names (by
formation-rule standards) of objects. The problem illustrates the
importance of keeping syntax and semantics distinct in defining a
formal language for logic.

In §30 Frege reduces the problem to showing that his simple names
denote by observing that his formation rules will preserve the property
of names being denoting. This is what we would expect in a proof by

induction on the complexity of formulas. In §31 he then turns to proving that his simple names denote.

Frege begins §31 by showing that his simplest primitive names denote. His simplest names, e.g., "$-\xi$", "$|\ \xi$", "$\mathsf{L}\xi$", "$\xi = \zeta$",

are all function-names. He first wishes to show that each of these denotes when a name denoting a truth-value is substituted for "ξ" and "ζ". He wants to do this because a function-name denotes only if upon completion by denoting object name(s) a denoting name results. But at the same time, an object-name denotes only if every completion of a function name by it results in a denoting name. These are the criteria set out in §29 which, though he disavows them as definitions of when his names denote, establish how we are to think and proceed in the proof. Their circularity infects the proof that follows. Frege's motivations are to be taken seriously; the disavowals are to keep us from thinking that the motivations play a role in the formal development of his Begriffsschrift.

Frege runs through these simplest primitive names and looks at what happens when names denoting the True or the False are substituted. Already we have a problem. To show why, let's compare his approach to a proof that every name in our modern first-order predicate calculus with functions has an interpretation in a given model. There we begin with the simplest names, show that they have interpretations, and proceed to complex ones, inducting on the complexity of their forms.[2] We can do this since there are simplest names that have interpretations set out by the giving of a model.

But Frege has a circular situation: his simplest names, e.g., "$\xi = \zeta$", have denotation if and only if every completion of them with denoting names results in a denoting name. There are no simple names that are given explicit denotations, except "$\varepsilon'(-\varepsilon)$" stipulated to denote the True, and "$\varepsilon'(\varepsilon = \mathsf{T}\cup\mathsf{J}\ \mathsf{a} = \mathsf{a})$" stipulated to denote the False.

At this point Frege isn't trying to show that just these names can be substituted for "ξ" and "ζ" in, say, "$\xi = \zeta$", as he doesn't even consider them separately until later in §31. Rather, he's talking about all names which denote truth-values, and these can be built up only through the use of primitive names. This is circular.

Using Frege's own example from the Appendix to Volume II of the *Die Grundgesetze der Arithmetik*, we can see how there is real circularity in his proof.[3] There Frege sets "$\forall$" to be an abbreviation of:

$$\varepsilon\,{}'\!\left(\left(\,\text{—}^{g}\!\!\cup\!\!\text{—}^{g\,(\varepsilon)}_{\;\;\;\varepsilon\,'(\,\text{—}g\,(\varepsilon)\,)=\varepsilon}\right)\right)$$

He then goes on to show that he can derive in his system both:

(1)
$$\text{—}^{g}\!\!\cup\!\!\text{—}^{g\,(\forall)}_{\;\;\;\varepsilon\,'(\,\text{—}g\,(\varepsilon)\,)=\forall}$$

and

$$\text{—}^{g}\!\!\cup\!\!\text{—}^{g\,(\forall)}_{\;\;\;\varepsilon\,'(\,\text{—}g\,(\varepsilon)\,)=\forall}$$

Now consider the course of values:

(2)
$$\alpha\,{}'\!\left(\left(\,\text{—}^{g}\!\!\cup\!\!\text{—}^{g\,(\alpha)}_{\;\;\;\varepsilon\,'(\,\text{—}g\,(\varepsilon)\,)=\alpha}\right)\right)$$

This is a primitive name by Frege's standards. He has already said on p. 87 (48) that "$\varepsilon'\,\phi(\varepsilon)$" is among the primitive names, and on p. 88 (49) of §31 he says,

> The matter is less simple with "$\varepsilon'\,\phi(\varepsilon)$"; for with this we are introducing not merely a new function-name, but simultaneously answering to every name of a first-level function of one argument, a new primitive name (course-of-values name); in fact not just for those [function-names] known already, but in advance for all such that may be introduced in the future.

That is, "$\varepsilon'\,\phi(\varepsilon)$" is a scheme of primitive names, so Frege must show that for every first-level ϕ which denotes, "$\varepsilon'\,\phi(\varepsilon)$" denotes. This is circular, as he's still supposed to be at the level of primitives. Moreover, he has to show that whatever result he gets by substituting any first-level function name for "ϕ" in "$\varepsilon'\,\phi(\varepsilon)$" when substituted in any of the primitive names (for example, "$\xi=\zeta$") results in a denoting name. This he must do in order to justify both that, for example, "$\xi=\zeta$" denotes and "$\varepsilon'\,\phi(\varepsilon)$" denotes for that substitution for "ϕ". Everything breaks down here—the example shows why: (1) doesn't denote, for if it did it would denote a truth-value, yet both it and its negation are derivable; hence (2) does not denote; hence the following, which is a substitution in "$\xi=\zeta$", does not denote:

(3) $(2) = \varepsilon'(\text{—}\varepsilon)$

So even "$\xi=\zeta$" does not denote.

Perhaps I'm reading this wrongly. Frege might say that on p. 88 (49) he doesn't have to show at the level of primitive names in his proof that (3) denotes. Why not? Because (2) doesn't denote, and that's because (1) doesn't denote. All he has to show is that for every completion of $"\xi = \zeta"$ by names that denote, a denoting name results. This appears to be the motivation of his definition of *fair (rechte)* course-of-values names on p. 88 (49). But this class is ill-defined, as it assumes we have a well-defined class of denoting first-level function-names, which is exactly what we are trying to establish.

The flaw in Frege's proof that all his names denote does not stem from the introduction of course-of-values. Even had he deleted names for those from his language, his proof as he attempted it would have been defective. With no simple denoting names he has no way to break into the circle of dependence of function-names and object-names on each other for their property of being denoting (see §29). Let me summarize why I think this problem occurs.

1. Being the first to formalize his language to this extent, Frege had no guide for how to prove something about the formalism. Today it seems obvious that to prove something like Frege wanted we'd proceed by induction on the complexity of the formation of names. But part of Frege's project was to carefully establish the principle of induction; hence he may have been unwilling to use that here. Or perhaps he didn't appreciate how induction should be carried out in this unusual situation, or his other conceptions led him astray, so he invokes properties of the whole class of formulas at the initial stage of induction/analysis, properties which he could claim only after he had established them.

2. The only simple object names in his language were course-of-values names, which included the truth-value names. But these are really complex names in that they depend on (not necessarily simple) function-names. So there was no obvious place to start an induction.

3. To show that his simple function-names denote, he had to say when a function-name denotes. How he perceived the answer to that is set out in §29, as described above. That view of when function-names denote comes from two more basic views:

> Functions are unsaturated, function-names need completion before they denote objects.

> Functions aren't themselves objects.

Frege's functions are so bloodless that he feels uneasy pointing to them and does so only indirectly by pointing at their courses-of-values. So he cannot begin his induction/analysis assuming or pointing to what his simple names denote.

4. Frege's courses-of-values are objects, but he can't switch to their names for the first step of his analysis, as noted in paragraph 2.

5. The now-standard way of inducting on the complexity of formation of names was blocked to Frege. He conceived of his course-of-values names as simple though they were complex in formation. His only way out was to assume that he had to induct on denoting names rather than all formulas.

6. That Frege chose to induct on denoting names may also be due to an ambiguity in setting out his formation rules. We'd say today that he was mixing his syntax and semantics. It's hard to follow whether he thought of any appropriate combination of signs as part of his Begriffsschrift: proper names are signs that are "supposed to denote an object" §26, p. 81 (44), which "are to denote something" p. 67 (32), and it's these that are involved in formation rules for function-names.

We could conjecture indefinitely on why Frege made the errors culminating in the incorrect proof of §30 and §31. But what we can say is that in the *Die Grundgesetze der Arithmetik* he tried but was unsuccessful in setting up a formal language, and that was due to placing semantic conditions on what counts as a correct formula.

Notes

1. (p. 150) All page references are to the English translation by M. Furth, followed in parentheses by the page reference to the original. No references are made to his earlier work *Begriffsschrift: A Formula Language, Modeled upon That of Arithmetic, for Pure Thought*.

2. (p. 151) See Chapter VIII of my *Classical Mathematical Logic*.

3. (p. 151) This is where Frege shows how to derive Bertrand Russell's paradox.

Postscript
Logic as the Art of Reasoning Well

In this last essay I'll try to summarize the conception of logic that
I have developed here and in the other volumes in this series: *The
Fundamentals of Argument Analysis*, *Prescriptive Reasoning*,
Reasoning in Science and Mathematics, and *Cause and Effect,
Conditionals, Explanations*. I'll do that by drawing two contrasts
that I see in work in logic, which I suspect are present in almost
all intellectual work.

Skeptic versus Dogmatist

I am a pyrrhonist. I am at a loss to understand what it would mean
to say that it's true that there are abstract objects in the world. I am
equally at a loss to understand what it means to say that it's true that
the world is made up of individual things, or that there are masses in
the world, or that there are no such things as properties, or that all is
one. There are so many ways to conceive of the world, and it makes no
sense to me to say that one of them is correct. We organize our understandings, we make our way through life with some assumptions that
we cannot test against the world because without those assumptions
we have no world. Mystics say that they have had experiences that
confirm one (or more) of these views. But such confirmation does not
survive translation into words that we can communicate one to another.

Still, we can compare these views. If you believe that the world is
made up of individual things, and you think that we can somehow pick
out or identify those things, then, with some further assumptions, we
can say that this is how you should reason. If you think the world is
process, with no individual things, no masses, just the one that is picked
out in parts with our language, then with some additional assumptions
we can set out a guide for how to reason well. We can compare, we
can try to see how our basic assumptions about the nature of the world
lead us to different methods and standards for reasoning. The basis
for logic, for all work in logic, is metaphysics. To the extent that we
can be clear about what we believe, we can bring those assumptions
about how to see the world into a form that we can share. We can

communicate our conceptions, and then, relative to those assumptions, our logics, our guides, have prescriptive force.

We investigate and compare differing views of the world through our logical studies, and we often revise our views as we develop our guides to reasoning. Metaphysics is the basis of logic, but as we develop our logic, we modify and understand better our metaphysics, most especially in comparing the outcome of different views. For example, second-order predicate logic, even second-order classical predicate logic, has many varieties as we add more or less to the metaphysics of first-order predicate logic. Developing methods for reasoning on the assumption that the world as process makes us see better how our usual assumption that the world is made up of individual things affects so much of how we talk and reason. A pyrrhonist is not a metaphysician. A pyrrhonist compares metaphysics, for that is all that he, that I, thinks we can do short of mysticism.

The assumptions that I start with in my work in logic are that there are people in the world, that we communicate through language, and that we believe our language in some way connects with the world so that we can classify some parts of our talk as true or false. Regardless of the view of the world you believe in, or adopt, or simply have, these assumptions seem necessary to reason together, which is how I justify adopting them.

Adding further assumptions we get a particular logic. Some, such as classical propositional logic, which starts with only these minimal assumptions, can serve those who hold very different views. Others, such as intuitionistic logic, make sense only relative to very specific assumptions about the world or the subject we are investigating. But to ask which is right seems to me to be a fruitless question, or at best a question that we can try to answer only by meditating or clearing our minds of all as we walk in the forest or sit by a stream, awaiting illumination (and hoping not to get struck by lightning).

I started long ago with the question "If logic is the right way to reason, why are there so many logics?". The answer I saw is that what logic you adopt depends on your metaphysics. But it also depends on what you are paying attention to. If you make only those minimal assumptions you get classical propositional logic. If you are worried about how we can come to know what is true, you might follow the intuitionists in their logic. If you are concerned with how likely it is that a claim is true, you might develop a many-valued logic.

When we come to investigate how to reason with prescriptive claims—about what should be done—the assumptions multiply and the contrasts are even greater. When we study how to reason about cause and effect, taking this comparative, skeptical view, we can devise methods that serve those who have greatly different conceptions, basic methods that can be made to serve more particular views of the subject by adding more assumptions and getting more particular methods.

Dogmatists disagree with this understanding of logic. The dogmatist has a particular metaphysics that he or she claims is right or true. They are certain that they have the insight into the nature of the world. For each—the platonist, the intuitionist, those who think that the world is made up of things and things only—there is just one correct logic. To the question "If logic is the right way to reason, why are there so many logics?" they can only point to how we may try to formulate logics that take into consideration more of the world than before. We pay attention not in our reasoning but in what we look at in the world, as a scientist. Other logics that are based on different metaphysics are just curiosities for them, not really logic at all, as a theory of physics in which there can be perpetual motion is just a curiosity.

Art versus Objective Science

In my work I strive to formulate methods for how to reason well. People matter. Logic is not a theory of the world, but a tool for us to develop theories of the world. Logic is not the end; the end is to try to find out what we are justified in believing. We use logic, it is an *organon*, a tool that is useful to all of us as we develop theories or try to decide which car to buy.

Arriving at beliefs that we can justify, however imperfectly, we cannot avoid judgment, the human, subjective aspect of logic. Yet it seems that there are at least some objective standards, particularly the criterion for an inference to be *valid*: it is impossible for the premises to be true and conclusion false. That, some say, depends on no judgment. Perhaps, though the discussion in "Valid Inferences and Possibilities" in this volume might make you doubt that, for what we count as a possibility depends on what we consider important in our reasoning. And, except for the simplest cases, recognizing that an inference is valid requires skill that seems like judgment.

With a valid inference the conclusion is certainly true—if the premises are certainly true. But only rarely do we begin with premises we know with certainty. We have to judge to what degree we are justified in believing them, how *plausible* they are. Even then, there are few opportunities in our lives to use valid inferences, as you can see in hundreds of examples in my *Critical Thinking*.

Rather, in our daily lives and almost all the time in the sciences we judge an inference to be good if it is *strong*: it is possible but not likely that the premises could be true and conclusion false. That "likely" is a judgment we make, relative to what we know, or at least relative to what we think we know and are paying attention to. It is a judgment that can be and often is revised as we learn more or as more of what we know is recalled to our attention. Using strong inferences to justify our beliefs, we have even less certainty about the conclusion that we have about the premises. But often, as I discuss in the other volumes in this series, and particularly in *The Fundamental of Argument Analysis*, that is all we want or need to live in our lives, apart and together.

For example, the following is clearly valid:

Ralph is a dog.
All dogs bark.
Therefore, Ralph barks.

But it is of no use in our reasoning because we know that "All dogs bark" is false. We use instead:

(*)   Ralph is a dog.
Almost all dogs bark.
Therefore, Ralph barks.

This is a strong argument that we can use in our reasoning. If we have good reason to believe "Ralph is a dog," then we have good reason to believe "Ralph barks." If we have good reason to believe that Ralph doesn't bark, then we have good reason to believe that "Ralph is a dog" is false. We conclude with less than certainty, but that is all we are likely to get in our lives, and that is enough for us to live our lives. We reason not without error. But who thinks that a life without error is possible or even desirable? Every step of our lives requires judgment.

But if logic involves such subjective evaluations, how is it different from psychology? We begin with our assumptions about the world. We make agreements about what we will pay attention to in our reasoning. Then relative to our agreements—which might be based on our seeing the world as it really is, or on how we as humans have no choice but to experience the world, or on choices we make to simplify and abstract in order to grasp some part of how we can evaluate our judgments—logic provides a standard, an *intersubjective* standard. Logic provides a prescription for how to reason well—relative to how we (choose to) see the world and what we (choose to) believe.

The pyrrhonist has none of this. The pyrrhonist accepts without justification only claims such as "I feel hot" or "I'm hungry" or "It seems to me that the sun is shining." In place of (*) the pyrrhonist has only:

> It seems to me that Ralph is a dog.
> It seems to me that almost all dogs bark.
> Therefore, it seems to me that Ralph barks.

This is clearly not valid. Much if not all of the work in these volumes is an attempt to replace that "it seems to me that" with intersubjective standards. Then logic is not psychology but a prescriptive art for how to reason well.

The human, the subjective is part of logic. In contrast, *objectivists* say that there is no place for the subjective in logic. Logic is the study of the world, perhaps the relations of abstract propositions and inferences or perhaps the most general truths of the world that we can formulate in our language. An objectivist need not be a dogmatist but can try to see how the world really is through developing different logics in order to compare them.

Nothing human, nothing subjective is part of the subject matter of logic for an objectivist. Much of what I have developed in these volumes would be classified by an objectivist not as logic but as philosophy of science, or philosophy of mathematics, or metaphysics, or meta-ethics, or the pragmatics of belief.

Objectivists typically take the scope of logic to be formal logic, the analysis of inferences for validity. But some wish to include more, perhaps all the subjects in the other volumes in this series. To do that, they hold that whether an inference is strong is not an intersubjective judgment but is objective, in the world independent of us. Whether a possibility is likely, they say, is a matter of objective probabilities. But as I show in *The Fundamentals of Argument Analysis*, that "likely" cannot be reduced to an analysis of objective probabilities, or if it could, it would leave us with no way to apply logic, for such probabilities are not knowable, not grounded in what we know or believe. The objectivist could reply that the latter doesn't matter, for that is how we apply logic, not what logic is.

For those who take logic to be an objective science, problems that arise in our everyday reasoning are not motive for developing new systems that have wider scope than the logics we already have. The only motive for devising a new logic would be to investigate objective relations that we have not previously considered. So, for example, to formulate a logic for prescriptive claims, the objectivist must assume that the truth-values and inferential relations of claims in ethics are independent of our judgments.

Logic as the art of reasoning well has wide scope: all the reasoning we want to do, all our search for justifying our beliefs and arriving at new beliefs through reasoning. It is fully human, involving judgment always. But it is not psychology; it is not the laws of thought. It is prescriptive: given this ordinary language, this view of the world that we codify in certain ways, paying attention to this, ignoring that aspect of language or experience, and given what we take as plausible claims, we have good reason to believe if we follow the guides of our logic. In contrast, logic as an objective science offers us no prescriptions, no help in our lives.

There is no right or wrong here. These are ways of dividing up our studies, leading to more or less fruitful and interesting work. But as for me, the choice is stark. We can restrict ourselves to only what is (seems to be) certain, what is (seems to be) independent of our human judgments and errors, and build a bulwark against the less than perfect in our lives. Or we can embrace the imprecision, the uncertainty, and try to reach out to each other to make agreements that can guide us in our reasoning and lives.

Come, let us reason together.

When people search for something, the likely outcome is that either they find it or, not finding it, they accept that it cannot be found, or they continue to search. So also in the case of what is sought in philosophy, I think, some people have claimed to have found the truth, others have asserted that it cannot be apprehended, and others are still searching. Those who think that they have found it are the Dogmatists, properly so called—for example, the followers of Aristotle and Epicurus, the Stoics, and certain others. The followers of Clitomachus and Carneades, as well as other Academics, have asserted that it cannot be apprehended. The Skeptics continue to search.

Sextus Empiricus *Outlines of Pyrrhonism,* 1.1
translated by Benson Mates in *The Skeptic Way*

Bibliography

Page references are to the most recent publication cited unless noted otherwise. *Italics* in quotations are in the original source.

ARISTOTLE
  1928 *De Interpretatione*
        Trans. E.M. Edghill, Oxford Clarendon Press. Also in *The Basic Works of Aristotle*, ed. Richard McKeon, Random House, 1941.
AZAR, Betty Schrampfer
  1989 *Understanding and Using English Grammar*
        2nd edition, Prentice-Hall.
BARWISE, Jon
  1975 *Admissible Sets and Structures*
        Springer-Verlag.
BAYS, Timothy
  2001 On Tarski on Models
        The Journal of Symbolic Logic, vol. 66, pp. 1701–1726.
BEALL, J. C. and Greg RESTALL
  2000 Logical Pluralism
        Australasian Journal of Philosophy, vol. 78, pp. 475–493.
BELL, J. L. and M. MACHOVER
  1977 *A Course in Mathematical Logic*
        North-Holland.
BENNETT, Jonathan
  1984 *A Study of Spinoza's Ethics*
        Hackett.
BERLIN, Isaiah
  1939 Verification
        Proceedings of the Aristotelian Society, vol. 39, pp. 225–248
        Reprinted in *The Theory of Meaning*, ed. G.H.R. Parkinson, Oxford University Press, 1978, pp. 15–34.
BOROWSKI, E. J.
  1974 Adverbials in Action Sentences
        Synthese, vol. 28, pp. 483–512.
BROADIE, Alexander
  1987 *Introduction to Medieval Logic*
        Oxford University Press.
CARNIELLI, Walter A., Marcelo E. CONIGLIO, Itala M. L. D'OTTAVIANO
  2009 New Dimensions on Translations between Logics
        Logica Universalis, vol. 3, pp. 1–18.

CARNIELLI, Walter A. and Itala M. L. D'OTTAVIANO
  1997   Translations between Logical Systems: A Manifesto
         Logique et Analyse, vol. 40, pp. 67–81.
CARNIELLI, Walter A. and Mamede LIMA-MARQUES
  1999   Society Semantics and Multiple-Valued Logics
         Advances in Contemporary Logic and Computer Scienc,e
         Number 235, pp. 33-52.
  See also EPSTEIN and CARNIELLI.
CASATI, Roberto and Achille VARZI
  2006   Events
         Stanford Encyclopedia of Philosophy online at
         <http://plato.stanford.edu/entries/events/>.
CORCORAN, John
  1980   Categoricity
         History and Philosophy of Logic, vol. 1, pp. 187–207.
DAVIDSON, Donald
  1967   Causal Relations
         Journal of Philosophy, vol. 64, pp. 691–703. Reprinted in
         Causation, eds. E. Sosa and M. Tooley, Oxford University Press,
         1993, pp. 75–87.
  1969   The Individuation of Events
         In *Essays in Honor of Carl G. Hempel*, ed. N. Rescher, D. Reidel.
DIAMOND, Jared
  2002   The Religious Success Story
         New York Review of Books, vol. 49, no. 17, November 7.
DUMMETT, Michael
  1964   Bringing About the Past
         The Philosophical Review, vol. 73, pp. 338–359. Reprinted in
         Dummett, 1977, pp. 333–350. References are to the first.
  1977   *Elements of Intuitionism*
         Clarendon Press.
EPSTEIN, Richard L.
  1990   *Propositional Logics (The Semantic Foundations of Logic)*
         Kluwer. 2nd edition, Oxford University, Press, 1995. 3rd edition,
         Advanced Reasoning Forum, 2012.
  1994   *Predicate Logic (The Semantic Foundations of Logic)*
         Oxford University Press. Reprinted Advanced Reasoning Forum, 2012.
  1998   *Critical Thinking*
         Wadsworth. 4th edition with Michael Rooney, Advanced Reasoning
         Forum, 2013.
  2005   Paraconsistent Logics with Simple Semantics
         Logique et Analyse, vol. 189–192, pp. 189–207.

2006   *Classical Mathematical Logic (The Semantic Foundations of Logic)*
       Princeton University Press.

2011   *Cause and Effect, Conditionals, Explanations*
       Advanced Reasoning Forum.

2012   *Reasoning in Science and Mathematics*
       Advanced Reasoning Forum.

2013A *The Fundamentals of Argument Analysis*
       Advanced Reasoning Forum.

2013B *Prescriptive Reasoning*
       Advanced Reasoning Forum.

20??   *Logic, Language, and the World*
       Volume 1: *The Internal Structure of Predicates and Names*
       Volume 2: *Time and Space in Formal Logic*
       Volume 3: *The World as Process*
         To appear, Advanced Reasoning Forum. A draft of all three
         volumes was previously posted on the website of the Advanced
         Reasoning Forum under the title *The Internal Structure of Pred-*
         cates and Names with an Analysis of Reasoning about Process.

201?   *Essays on Language and the World*
       In preparation.

EPSTEIN, Richard L. and Esperanza BUITRAGO-DÍAZ

2014   A Proposition Logic of Temporal Connectives
       Logic and Logical Philosophy, vol. 24, no. 2, pp. 155–200.
       DOI 10.12775/LLP.2014.015, 2014.

EPSTEIN, Richard L. and Walter A. CARNIELLI

1989   *Computability*
       Wadsworth & Brooks/Cole. 3rd edition, Advanced Reasoning
       Forum, 2008.

EPSTEIN, Richard L. and Itala M. L. D'OTTAVIANO

1988   A Paraconsistent Many-Valued Logic: J_3
       Reports on Mathematical Logic, vol. 22, pp. 89–103.

EPSTEIN, Richard L. and Stanislaw KRAJEWSKI

2004   Relatedness Predicate Logic
       Bulletin of Advanced Reasoning and Knowledge, vol. 2, 2004,
       pp. 19–38. Available at <www.AdvancedReasoningForum.org>.)

EPSTEIN, Richard L., Fred KROON, and William S. ROBINSON

2013   Subjective Claims
       In EPSTEIN, 2013A, pp. 95–130.

ETCHEMENDY, John

1990   *The Concept of Logical Consequence*
       Harvard University Press.

FREGE, Gottlob
    1893   *Die Grundgesetze der Arithmetik*
            Translated as *The Basic Laws of Arithmetic* by M. Furth, University
            of California Press, 1967.
    1918   Der Gedanke: eine logische Untersuchung
            Beträge zur Philosophie des deutschen Idealismus, pp.58–77. Trans.
            by A. and M. Quinton as "The Thought: A Logical Inquiry" in *Mind*,
            (new series) vol.65, pp.289–311, reprinted in *Philosophical Logic*,
            ed. P. F. Strawson, Oxford University Press, 1967, pp.17–38.
    1919   Negation
            Beiträge zur Philosophie des deutschen Idealismus, pp. 143–157.
            Trans. by P. T. Geach in *Translations from the Philosophical
            Writings of Gottlob Frege*, Basil Blackwell, 1970, pp. 117–135.
GOMBRICH, E.H.
    1961   *Art and Illusion*
            2nd edition, Princeton University Press.
GOULD, James L. and Carol Grant GOULD
    1994   *The Animal Mind*
            Scientific American Library.
HAAS, W.
    1962   The Theory of Translation
            Philosophy, vol. 37, no. 141, pp. 208–228.
HANSON, William
    1997   The Concept of Logical Consequence
            The Philosophical Review, vol. 106, 1997, pp. 365–409.
HUGHES, G.E.
    1982   *John Buridan on Self-Reference*
            Cambridge University Press.
ISAACSON, Daniel
    1996   Arithmetical Truth and Hidden Higher-Order Concepts
            In *The Philosophy of Mathematics,* ed. W. D. Hart, Oxford
            University Press.
ISLES, David
    1981   Remarks on the Notion of Standard Non-Isomorphic
            Natural Number Series
            In *Constructive Mathematics,* ed. F. Richman, Springer-Verlag
            Lecture Notes in Mathematics, no. 873, pp. 111–134.
KADVANY, John
    1989   Reflections on the Legacy of Kurt Gödel: Mathematics, Skepticism,
            Postmodernism
            The Philosophical Forum, vol. 20, pp. 161–181.

KRAJEWSKI, Stanislaw
  1986   Relatedness logic
         Reports on Mathematical Logic, vol. 20, pp. 7–14.
  2003   *Twierdzenie Gödla i Jego Interpretacje Filozoficzne:*
         od Mechanicyzmu do Postmodernizmu
         Wydawnictwo IFiSPAN, Warszawa.
  2003   Gödel's Theorem and Its Philosophical Interpretations:
         From Mechanism to Postmodernism
         Bulletin of Advanced Reasoning and Knowledge, vol. 2,
         pp. 103–108. Also at <www.AdvancedReasoningForum.org>.
         See also EPSTEIN and KRAJEWSKI.
KRIPKE, Saul
  1976   Is There a Problem about Substitutional Quantification?
         In *Truth and Meaning,* eds. Gareth Evans and John McDowell,
         Oxford University Press.
LEBLANC, Hugues
  1973   Semantic Derivations
         In *Truth, Syntax and Modality,* ed. H. Leblanc, North-Holland.
LE POIDEVIN, Robin
  1990   Relatonism and Temporal Topology
         Philosophical Quarterly, vol. 40, pp. 419–432. Reprinted in
         Le Poidevin and MacBeath, *The Philosophy of Time,* Oxford
         University Press, 1993, pp. 149–167.
LEWIS, C. I. and C. H. LANGFORD
  1932   *Symbolic Logic*
         The Century Company. 2nd ed. with corrections, Dover, 1959.
LYNCH, Michael P.
  2009   *Truth as One and Many*
         Oxford University Press.
LYONS, John
  1968   *Introduction to Theoretical Linguistics*
         Cambridge University Press.
MATES, Benson
  1953   *Stoic Logic*
         University of California Press. Reprinted 2014, Advanced Reasoning
         Forum.
MELLOR, D. H.
  1982   Theoretically Structured Time: A Review of W. H. Newton-Smith
         The Structure of Time
         Philosophical Books, vol. 23, 1982, pp. 65–69.
         See also NEWTON-SMITH, 1982.

MOSSAKOWSKI, Till, Razvan DIACONESCU, and Andrzej TARLECKI
  2009   What is a Logic Translation?
         Logica Universalis, vol. 3, 2009, pp. 95–24.
MOSTOWSKI, Andrzej
  1951   On the Rules of Proof in the Pure Functional Calculus of the
         First-Order
         Journal of Symbolic Logic, vol. 16, pp. 107–111.
NAESS, Arne (NESS)
  1938   *"Truth" as Conceived by Those Who Are Not Professional
         Philosophers*
         Skrifter utgitt av Det Norske Videnskaps-Akademi i Oslo, II. Hist.–
         Filos. Klasse, No. 4. Reprinted by the Advanced Reasoning Forum, 2014.
NEEDHAM, Paul
  2011   Micoressentialism: What is the Argument?
         Noûs, vol. 45, pp. 1–21.
NEWTON-SMITH, W. H.
  1982   Reply to Dr. Mellor
         Philosophical Books, vol. 23, 1982, pp. 69–71.
         See also MELLOR, 1982.
NOZICK, Robert
  1981   Why Is There Something rather than Nothing?
         In *Philosophical Explanations*, Harvard University Press.
PARSONS, Terence
  1990   *Events in the Semantics of English*
         The MIT Press.
PELLETIER, Francis Jeffrey and Lenhart K. SCHUBERT
  1989   Mass Expressions
         In Volume IV of *Handbook of Philosophical Logic*, eds. D. Gabbay
         and F. Guenthner, Reidel, pp. 327–407.
PLANTINGA, Alvin
  1974   *The Nature of Necessity*
         Oxford University Press.
PRIOR, Arthur
  1957   *Time and Modality*
         Oxford University Press.
  1962   Changes in Events and Changes in Things
         Preprint. Reprinted in *The Philosophy of Time*, eds. Robin Le
         Poidevin and Murray MacBeath, Oxford University Press, 1993,
         pp. 35-46. Also reprinted in *Papers on Time and Tense* by Arthur
         Prior, Oxford University Press, 2003, pp. 7–19.
  1967   *Past, Present and Future*
         Oxford University Press.

PUTNAM, Hilary
 1973 Meaning and Reference
    The Journal of Philosophy, vol. 70, pp. 699–711.
 1975 What Is Mathematical Truth?
    In *Mathematics, Matter, and Method: Philosophical Papers, vol. 1*,
    Cambridge University Press, 2nd edition, 1979, pp. 60–78.
QUINE, Willard Van Orman
 1939 Designation and Existence
    The Journal of Philosophy, vol. 36, no. 26, pp. 701–709.
 1950 *Methods of Logic*
    Harvard University Press. 4th ed., 1982.
 1953 On What There Is
    In *From a Logical Point of View*, W. V. O. Quine, Harvard
    University Press, 2nd ed., 1961, pp. 1–19.
RAY, Greg
 1996 Logical Consequence: A Defense of Tarski
    Journal of Philosophical Logic, vol. 25, pp. 617–677.
READ, Stephen
 1994 Formal and Material Consequence
    Journal of Philosophical Logic, vol. 23, pp. 247–265.
RESCHER, Nicholas and Alasdair URQUHART
 1971 *Temporal Logic*
    Springer-Verlag.
ROBINSON, Julia
 1949 Definability and Decision Problems in Arithmetic
    The Journal of Symbolic Logic, vol. 14, pp. 98–114.
SAGÜILLO, José M.
 1997 Logical Consequence Revisited
    The Bulletin of Symbolic Logic, vol. 3, pp. 216–241.
SEARLE, John R.
 1983 *Intentionality*
    Cambridge University Press.
SHER, Gila
 1996 Did Tarski Commit "Tarski's Fallacy"?
    The Journal of Symbolic Logic, vol. 61, pp. 653–686.
SMILEY, T. J.
 1976 Comment on 'Does Many-Valued Logic Have Any Use?' by D. Scott,
    In *Philosophy of Logic*, ed. S. Körner, Univ. of California, pp. 74–88.
SZCZERBA, Lesław
 1977 Interpretability of Elementary Theories
    In *Logic, Foundations of Mathematics and Computability Theory*,
    eds. R. I. Butts and J. Hintikka, D. Reidel, pp. 129–145.

TARSKI, Alfred
  1956   On the Concept of Logical Consequence
         In Tarski, *Logic, Semantics, Metamathematics*, Oxford
         University Press.
UNWIN, Nicholas
  1996   The Individuation of Events
         Mind, vol. 105, pp. 315–330.
WAISMANN, Friedrich
  1945   Verifiability
         Proceedings of the Aristotelian Society, Supp. Vol. 19, 1945,
         pp. 119–150. Reprinted in *The Theory of Meaning* ed. G. H. R.
         Parkinson, Oxford University Press, 1968, pp. 33–60.
  1968   How I See Philosophy
         In Waismann, *How I See Philosophy*, Macmillan, pp. 1–38.
WANG, Hao
  1996   *A Logical Journey: From Gödel to Philosophy*
         MIT Press.
WHITE, Alan R.
  1970   *Truth*
         Anchor Books, Doubleday & Company.
WHORF, Benjamin Lee
  1940   Science and Linguistics
         Technology Review vol. 42, pp. 229–231, 247–248. Reprinted in
         Language, Thought, and Reality : *Selected writings of Benjamin
         Lee Whorf*, ed. John B. Carroll, The MIT Press, 1956, pp. 207–219.
WILLIAMSON, Colwyn
  1968   Propositions and Abstract Propositions
         In *Studies in Logical Theory*, ed. N. Rescher, *American
         Philosophical Quarterly,* Monograph no. 2, Basil Blackwell, Oxford.
WRIGHT, Cory D.
  2005   On the Functionalization of Pluralist Approaches to Truth
         Synthese, vol. 145, pp. 1–28.
WRIGLEY, Michael
  1980   Wittgenstein on Inconsistency
         Philosophy, vol. 55, pp. 471–484.
YOUNG, James O.
  2008   The Coherence Theory of Truth
         The Stanford Encyclopedia of Philosophy, <http://plato.stanford.
         edu/entries/truth-coherence/>, accessed on 8/8/12.

Index of Symbols

Semantics

Index

italic page numbers indicate
a definition or a quotation

CPSIA information can be obtained
at www.ICGtesting.com
Printed in the USA
BVOW04s1307090817
491601BV00016B/96/P